The Drum Motor

Stefan Hamacher

The Drum Motor

The All-Rounder in Modern Unit Handling
Conveyor Technology

Stefan Hamacher
Interroll Trommelmotoren GmbH
Hückelhoven, Germany

ISBN 978-3-662-59297-7   ISBN 978-3-662-59298-4 (eBook)
https://doi.org/10.1007/978-3-662-59298-4

Springer Vieweg

This Vieweg imprint is published by the registered company Springer-Verlag GmbH, DE, part of Springer Nature.
The registered company address is: Heidelberger Platz 3, 14197 Berlin, Germany

Table of Contents

Foreword

The majority of people use drum motors nearly every day without being aware of it.

It might be at the supermarket checkout, while checking in for a flight or during the hand luggage security check at the airport. Most of our food is produced safely and hygienically with the aid of drum motors; so, among other things, the drum motor has also contributed to the fact that the quality of industrially manufactured food has improved enormously in recent decades.

Although the drum motor may be found everywhere, it is practically invisible, as it is perfectly adapted to and integrated with its environment. Therefore, most people do not even see it at all. But there really is no reason for the drum motor to shy away from the spotlight. In addition to its adaptability, it is also watertight and maintenance-free. These are only three of the many features that make the drum motor so unique and why more and more conveyor belt manufacturers are opting for a drive concept with an integrated drum motor.

Feedback and suggestions from users are always gratefully received. The quickest way to contact me is via email: s.hamacher@interroll.com

In closing, I would like to thank Interroll for their generous support towards making this book a reality.

Selfkant, March 2019   Stefan Hamacher

Structure of conventional conveyors 1

Conventionally, drum motors are used as drives in conveyor systems. The following chapters explain the most important terms and designs in conveyor belt engineering. They will be used again and again throughout this book.

For the most part, the focus is on intralogistics applications for the transport of unit loads (see Fig. 1.1), including their dimensioning in chapter 5.

The book does not cover special considerations regarding the dimensioning and construction of heavy bulk material conveyors (see Fig. 1.2) such as those found for example in open-cast mining or underground.

Unit loads include products that are transported in defined units, e.g. boxes, pallets, containers, larger parts and objects.

Bulk material refers to granular goods and material that is broken down into small parts making it pourable, bulk cargo, e.g. sand, gravel, ore, coal, grain, salt, sugar, coffee, flour.

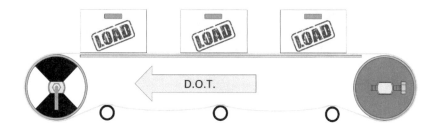

Fig. 1.1 Unit load handling

© Springer-Verlag GmbH Germany, part of Springer Nature 2020
S. Hamacher, *The Drum Motor*,
https://doi.org/10.1007/978-3-662-59298-4_1

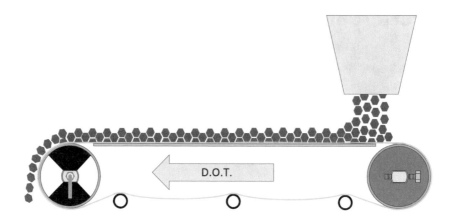

Fig. 1.2 Bulk material conveyor

1.1   Structure of conventional conveyor belts

A conventional conveyor for unit load handling consists essentially of a drum motor or a drive drum with external motor, a idler pulley, a conveyor belt and a slider or roller bed. (See Fig. 1.3)

Depending on requirements, snub, deflection, tensioning and support rollers may also be used.

The belt tension in friction-driven conveyor belts can be adjusted by means of tensioning screws on the pulley.

The belt tension is necessary to create grip between the conveyor belt and drum.

The tensioning screw on the drum motor or drive drum is required for fine tuning the running of the belt.

Definition of terms:

EL: Effective length of a drum motor or roller between conveyor frames
SL: Shell length of a drum motor or a roller
TM: Drum motor or drive drum
UT: Idler pulley
A-A: Axis-to-axis length between drum motor/drive drum and pulley
BW: Conveyor belt width
SW: Tensioning screw
DOT: Direction of travel
◗◖:   Rotating driveInlineshape

The top face of a conveyor is called the upper or carrying strand.
The conveyor belt is usually run along a slider bed or roller bed on the upper strand.
In most cases, the goods to be conveyed are carried on the upper strand.

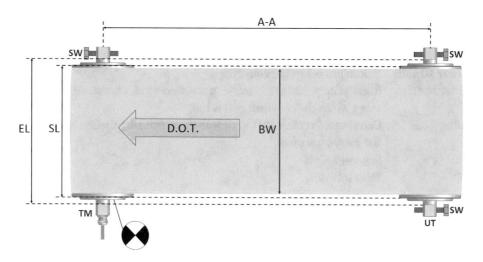

Fig. 1.3 Top view of a belt conveyor

Most conveyors in the unit load handling sector are constructed with a slider bed, as this type of conveying surface is relatively simple and usually the most cost-effective.

However, this comes at the expense of the conveyor's efficiency, because a conveyor belt that has a slider bed requires more energy than, for example, a conveyor with a roller bed on account of greater frictional losses.

The conveyor belt is returned to the pulley along the lower (or return) strand.

Snub rollers ensure a belt wrap angle of 180°–270° on the drum motor or drive drum.

Support rollers on the lower strand serve to lift the conveyor belt, so that it cannot sag too much.

As a rule, the conveyor belt is connected in a continuous, closed loop.

A conveyor belt is therefore often also defined by its endless length.

The endless length of a conveyor belt is the length of the closed conveyor belt measured from the connection seam and ending there again after one revolution.

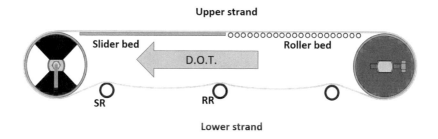

Fig. 1.4 Side view of a belt conveyor

Definition of terms:

Upper strand: Top face of the conveyor and the side on which the cargo is usually transported.

Lower strand: Belt return below the conveyor

Slider bed: Rigid plate or slide rails made of stainless steel, plastic, steel, wood … along which the conveyor belt is run.

Roller bed: Carrying rollers replace the slider bed, minimising friction between belt and carrying surface.

RR: Support rollers

SR: Snub rollers

The type of conveyor is determined by the position at which the drum motor or the drive drum is placed in the conveyor system.

There are conveyors with head drive, centre drive or tail drive.

If one or more drum motors or drive drums are used to drive a roller train, this is referred to as a roller conveyor.

1.2   Conveyor belts with head drive

The most common type of conveyors in unit load handling are conveyors with a head drive. (See Fig. 1.5)

If the drum motor or drive drum is installed at the head of the conveyor belt, i.e. in the direction of travel, the conveyor belt is pulled steadily, so the upper strand always remains taut. Head-driven conveyors are frequently operated in one direction only.

The head drive is ideal for ascending conveyor belts (see Fig. 1.6), as the drum motor or drive drum can pull the load upwards in this application.

For steeper gradients, a backstop or holding brake should be installed in the drum motor or on the drive drum to prevent uncontrolled running back of the loaded conveyor belt after shutdown or power failure.

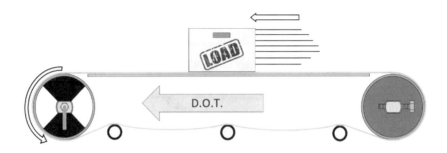

Fig. 1.5 Conveyor belt with head drive

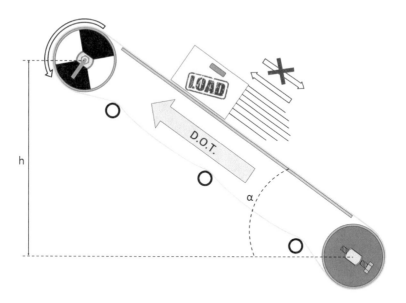

Fig. 1.6 Ascending conveyor with head drive

The head drive ensures that the conveyor belt in an ascending conveyor always remains taut, even if the belt should have to stop.

The conveyor drive has to work harder in an ascending conveyor than in a comparable horizontal conveyor owing to the additional force of gravity acting in opposition to the belt tension of the conveyor drive.

The steeper the angle of inclination, the more driving power is needed.

The inclination of a conveyor must therefore be taken into account when dimensioning the drive.

If a horizontal conveyor belt with head drive is operated in the opposite direction of travel it is referred to as a tail-driven conveyor. The drum motor thus pushes the conveyor belt.

Particularly in conveyors in which the conveyor belt hangs down loosely in the lower strand, the conveyor belt can bunch together behind the load on the upper strand with the result that a "belt bulge" is pushed along ahead of the load.

The smooth running of the belt can be affected and in the worst case cause the belt to run uncontrollably.

1.3   Conveyor belts with centre drive

Conveyors with a centre drive are particularly suitable for keeping the conveyor belt constantly taut, even in applications where bidirectional rotation in the upper strand is required (see Fig. 1.7).

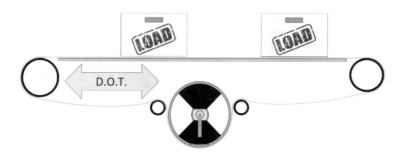

Fig. 1.7 Conveyor belt with centre drive

The drum motor or drive drum is positioned in the lower strand.

The conveyor belt is laid around the drum motor or the drive drum using snub rollers to achieve a wrapping angle of approx. 180°–270°.

This design is somewhat more elaborate than the head-driven conveyor belt, but offers a range of advantages for several applications.

A conveyor belt design with centre drive allows for the use of smaller pulleys, thus enabling smaller transfers as well.

This is particularly an advantage if the material being conveyed has to be transferred from one conveyor belt to another.

If the pulleys of two conveyor belts are too large where a load has to be transferred from one conveyor to the other the load can get stuck at the point of transfer. (See Fig. 1.8)

Fig. 1.8 Problem caused by too large a transfer gap between two conveyor belts

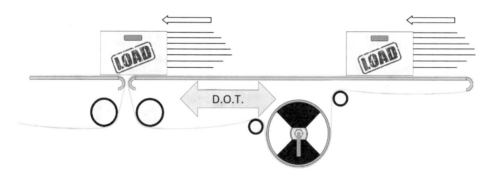

Fig. 1.9 Knife edges

1.3.1 Knife edges

The smallest possible deflection or transfer can be achieved with so-called knife edges. (See Fig. 1.9)

There are rigid and rolling knife edges.

A rigid knife edge is a rounded edge or a thin rod, around which the conveyor belt is very tightly wrapped.

However, the extremely small deflection over a rigid knife edge generates additional frictional heat, which can cause the knife edge and the conveyor belt to heat up considerably.

Replacing the rigid knife edge with a rolling knife edge can reduce the frictional losses.

A rolling knife edge can be a small ball bearing roller, for example.

The small roller must, however, be strong enough to withstand the tension in the belt.

Broad knife edges can therefore only be implemented with great difficulty for rolling transfers.

Frictional losses can be reduced by reducing the belt wrap angle at the knife edge.

1.4 Conveyor belts with tail drive

Conveyor belts are rarely designed with a tail drive.

A tail drive conveyor is, in principle, a head-driven conveyor operating in the reverse direction. (See Fig. 1.10)

In order to prevent the conveyor belt bunching up behind the transported load as far as possible, greater belt tension is required in a tail drive conveyor, so that the conveyor belt in the lower strand is hardly able to sag at all.

Since the conveyor belt is now pushed along the upper strand and not pulled, the full torque of the drum motor or drive drum may no longer be delivered to the conveyor belt due to a lack of grip between conveyor belt and drum.

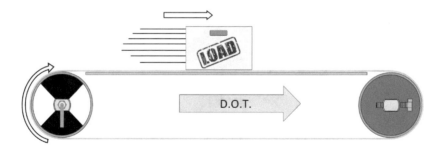

Fig 1.10 Conveyor belt with tail drive

Tail-driven conveyors are suitable for short conveyors with light loads or for descending conveyors. (See Fig. 1.11)

A descending conveyor can also be used in the opposite direction of travel as an ascending conveyor.

As a general rule, the drive should always be placed at the highest point on ascending or descending conveyors.

With the descending conveyor, the transported load tightens the conveyor belt in the upper strand and excess conveyor belt is pushed down to the lower strand ahead of the load. If the gradient is too steep, the drum motor or drive drum should be equipped with a holding brake.

A holding brake is usually open while powered, so the drum motor or drive drum can rotate freely.

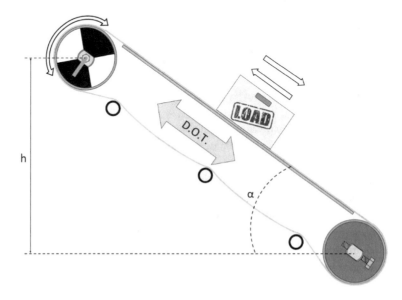

Fig. 1.11 Descending conveyor with tail drive

If the holding brake is switched off or if there is a power failure, then the holding brake closes automatically, preventing an uncontrolled discharge of the load.

A holding brake must be switched on or off simultaneously with the power supply of the drive. The simultaneous switching can be achieved by means of relays or contactors.

If the drive were to run when the holding brake is closed, then the motor coil or the holding brake could get damaged.

A backstop does not make sense in a descending conveyor, as it would have to be freely rotating in the direction of travel, i.e. downwards.

1.5   Roller conveyors

A roller conveyor usually consists of several successively arranged rollers. (See Fig. 1.12 and 1.13)

The actively driven rollers are called "master rollers", the passively driven rollers are called "slave rollers".

Several drives can be installed in a roller conveyor, which need to be aligned with each other in terms of speed.

The drum motors or drive drums are connected directly to the "slave rollers" with chains, toothed belts or other transmission media so that the "slave rollers" are also driven.

The rollers are *not* mechanically connected between two drive zones.

Roller conveyors can generally be operated bidirectionally.

However, they are not suitable for inclined conveyor lines. Descending conveyor lines usually do not use drive motors but rather take advantage of gravity.

Roller conveyors are extremely energy efficient due to their low rolling friction.

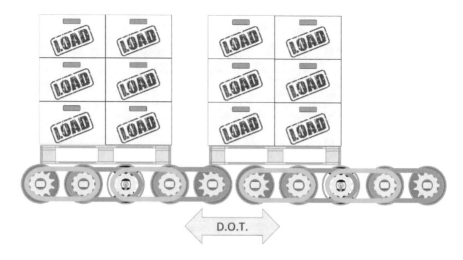

Fig. 1.12 Roller conveyor for transporting pallets (side view)

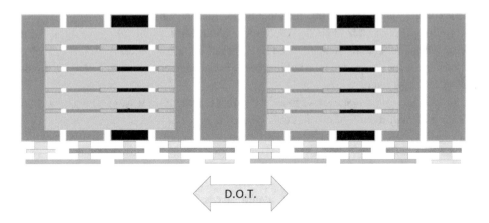

Fig. 1.13 Roller conveyor for transporting pallets (top view)

It is possible to transport many times more load mass with a roller conveyor than with a belt conveyor that has a sliding conveying surface at the same drive power.

Roller conveyors are suitable for larger products with a flat lower surface such as pallets, crates, boxes, beams, steel girders …

If the product to be conveyed is too small for the roller conveyor, there is a risk that the load will get stuck between the individual rollers or even fall through.

Types of conveyor belts

2

The conveyor belt has a direct influence on the behaviour and lifespan of the drum motor or drive drum.

It is *crucial* to recognise and understand the extreme effects of the conveyor belt on the drum motor when dimensioning a conveyor drive.

The choice of conveyor belt usually depends on the application.

There are many different types of conveyor belts and drive systems. The most common conveyor belts and drive systems are described in the chapters that follow.

2.1 Conventional friction-driven flat belts

Conventional friction-driven flat belts are widely used. (See Fig. 2.1) These conveyor belts come in a broad range of designs and materials.

With all of these belts, the power of the drum motor or drive drum is transmitted by friction. The difficulty with friction-driven flat belts lies in building up enough friction or grip between the conveyor belt and the drum without mechanically overstressing the drum motor or drive drum.

The grip between the drum and the belt is usually generated through belt tension.

This is achieved by mechanically pressing the conveyor belt against the drum.

The belt tension may not be too great, otherwise the ball bearings of the drum motor or the bearings of the drive drum can be damaged by the high belt tension forces.

The drum shell is often also knurled or has a rubber lagging to improve the frictional engagement or grip.

Care should be taken with knurled drum shells, however, as the knurls are sometimes sharp and can damage the underside of the conveyor belt under heavy load.

© Springer-Verlag GmbH Germany, part of Springer Nature 2020
S. Hamacher, *The Drum Motor*,
https://doi.org/10.1007/978-3-662-59298-4_2

Fig. 2.1 Drum motor with friction-driven flat belt (Source: Interroll.com)

Increasing the wrap angle of the conveyor belt to more than 180° further improves the frictional engagement between the drum and the belt as a result of the larger contact surface.

Tracking a friction-driven flat belt i.e. keeping it running straight, is usually achieved by using a crowned drum shape on the drive side.

For short, almost square, conveyors with head drives the pulley is often cylindrical.

For belts where the axle-to-axle distance (A-A) is more than 5 times longer than the belt width, or belts that are head-driven, the pulley should be crowned if at all possible.

The belt tension is highest in the centre of a crowned drum, because the conveyor belt stretches the most in the middle due to the bulge of the crown.

This creates a force which centres the conveyor belt in the middle of the drum.

The belt tension is usually adjusted around the pulley.

The belt tracking should only be aligned on the drum motor or drive drum.

The specifications of the conveyor belt are extremely important for dimensioning the drive.

If the conveyor belt is not known, you cannot perform any serious drive calculations.

There is usually a data sheet available for every conveyor belt, which can be found on the internet or can be requested from the belt manufacturer.

The following belt details are important for drive dimensioning:
BW: Conveyor belt width
The width of the conveyor belt should be easy to determine. If necessary, it can simply be measured with a tape measure.

Bt: Conveyor belt thickness
The thickness of the conveyor belt does influence the final belt speed or the unwind roll diameter of the belt, although in most cases this influence is minimal.

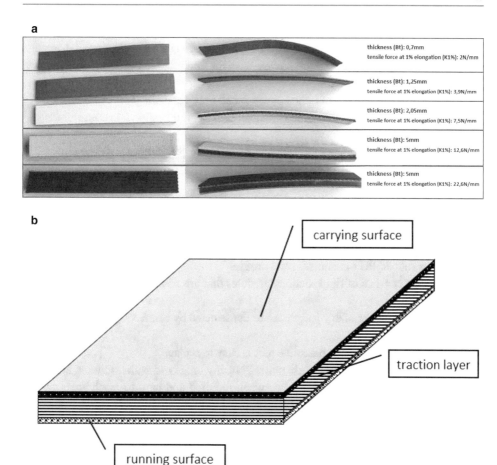

Fig. 2.2 a A range of fabric belts. **b** Example structure of a fabric belt

K1%:   Tensile force [N] per millimetre [mm] of belt width at 1% elongation. [N/mm]
This value is extremely important. It can be used to calculate the force acting on the ball bearings of the drum motor or drive drum due to belt tension. The K1% value indicates how much tensile force in [N] per 1 mm conveyor belt width acts on the drum motor and pulley when the belt is stretched to 1% elongation.

mb: Belt mass [kg/m^2]
The mass of the conveyor belt can be calculated using the conveyor belt specific mass. After all, the belt mass must also be moved in addition to the load.

Fabric belts usually consist of different layers. Depending on the structure and materials used, the belts may be stretchy or more rigid, with varying K1% values. (See Fig. 2.2 a and b)

Each conveyor belt has a minimum deflection diameter that depends on its structure and composition. When selecting the conveyor belt, it must be ensured that the outer diameters of the drum motor and the pulley are not less than the minimum deflection diameter of the conveyor belt.

If the drum diameter is too small, the conveyor belt may get damaged.

Having too small a deflection diameter can also lead to high frictional losses.

2.2   Modular positive drive belts

Modular positive drive belts are very common in food applications, such as for meat processing in wholesale butcheries. (See Fig. 2.3)

Power is transmitted from the drum shell to the modular belt via interlocking sprockets.

The positive locking can be achieved through the use of sprocket wheels or by means of a continuous rubber, PU or stainless steel profile.

A modular belt consists of rigid sections/modules that are connected to a hinge rod. (See Fig. 2.4)

In this way, endless belts of any length can be constructed by connecting as many modules as desired.

If a module gets damaged, the defective part is easy to replace.

The modular belt is opened by simply pulling out any hinge rod with a suitable tool.

The modular belt is thus easy to remove when required and hard-to-reach areas can be maintained more easily.

Owing to the positive power transmission, no belt tension is required. The ball bearings of the drum motor or drive drum are thus under less strain, which greatly reduces their probability of failure.

The positive power transmission always continues to drive the modular belt reliably in very wet applications. There is generally no slipping of the belt with modular positive drive belts, as may occur, for example, in a friction-driven conveyor belt.

The individual modules are made of PE, PP or POM among other materials and are very stable and solid. As a result, most modular belts are usually cut-resistant.

Examples of cut-resistant modular belts are found in carving belts, where the product often has to be processed on the conveyor belt using sharp knives.

Modular belts are 5 to 15 times heavier than comparable friction-driven flat belts due to their greater bulk.

The mass of the belt must therefore be taken into account when dimensioning the drive.

The coefficient of friction between the modular belt and the slider bed may vary depending on the material that the modular belt is made of, but it is often slightly lower than with friction-driven conveyor belts.

Fig. 2.3 Modular belt application in the food industry (Source: Interroll.com)

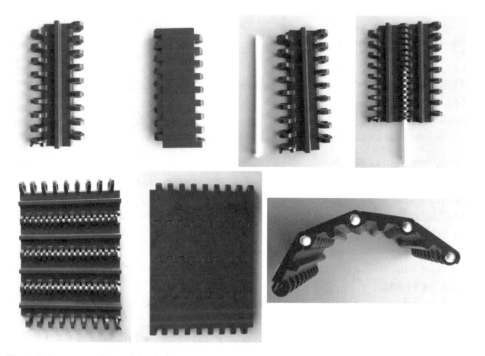

Fig. 2.4 Structure of a modular belt

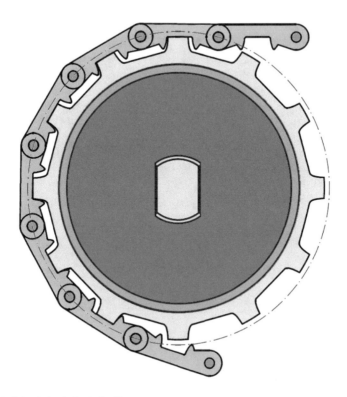

Fig. 2.5 Modular belt pitch circle diameter

The following modular belt details are important for dimensioning the drive (see Fig. 2.5):

mb: Belt mass [kg/m²]

The mass of the modular belt can be calculated using the conveyor belt specific mass.

PCD: Pitch circle diameter

The pitch circle diameter is the diameter of an imaginary circle passing through the centre of the modular belt's hinges.

The final belt speed or the unwind roll diameter of the module belt can be determined using the pitch circle diameter.

There is a very wide range of modular belts available. (See Fig. 2.6)

Modular belts with a pronounced side profile can be driven by continuous profiles or sprocket wheels.

Modular belts with a side profile that resembles that of a bicycle chain can only be driven, like a bicycle chain, by sprocket wheels.

The number of sprocket wheels required depends on the load and the width of the modular belt and it should be calculated or given by the belt manufacturer for each application.

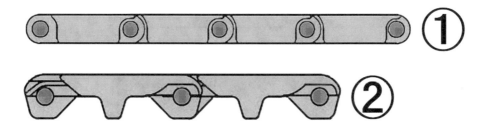

Fig. 2.6 Modular belt side profiles

As a rough rule of thumb, one can assume, however, that one wheel is needed per 100 mm of modular belt width.

Explanation of Fig. 2.6:

① Modular belt with lateral profile similar to a bicycle chain.
   This modular belt can only be driven by means of sprocket wheels.
② Modular belt with pronounced side profile. The teeth on the underside of the belt can interlock with a continuous shell profile. Alternatively, this type of modular belt can also be used with sprocket wheels. (See Fig. 2.7)

The power transmission between drum shell and sprocket wheel is often effected via one or more steel wedges welded onto the drum shell.

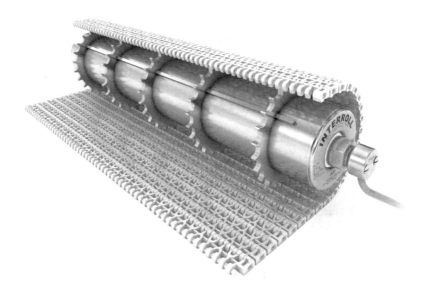

Fig. 2.7 Drum motor with sprocket wheels (Source: Interroll.com)

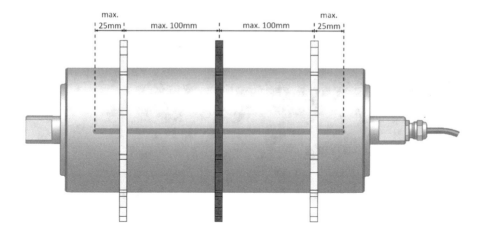

Fig. 2.8 Modular belt tracking with a fixed sprocket *wheel (red)*

The sprocket wheels thus have a groove and are pushed over the steel wedge on the drum shell. The sprocket wheels must be able to move freely.

Theoretically, one could fix all the sprocket wheels onto the drum shell. However, one would have to align the wheels very precisely, so that the teeth can properly engage with the modular belt without damaging it.

Loose sprocket wheels eliminate the risk that the teeth will be misaligned because the loose sprocket wheels can self-align with the modular belt.

In addition, having loose sprocket wheels facilitates the cleaning of a modular belt application, because they can be easily pushed aside for cleaning purposes.

However, it is important to ensure that the sprocket wheels are pushed back to the correct places after cleaning.

A modular belt can be guided by means of lateral guide rails, which hold the modular belt in position on the left and right.

An alternative to lateral belt guidance is to fix the middle sprocket wheel in position for tracking purposes. (See Fig. 2.8)

There are sprocket wheels that can be fastened to the drum shell with screws for this reason.

The total number of sprocket wheels should be odd for modular belt tracking by means of a centrally fixed sprocket wheel.

All the other sprocket wheels are then slid loosely onto the drum shell to the left and right. Sprocket wheels have the advantage that you can replace them as needed.

This allows the user to easily change the modular belt series or type if necessary.

However, sprocket wheels made of stainless steel or other hard materials can generate rattling noises.

An application that has several modular belt conveyors can sometimes build up a high level of background noise.

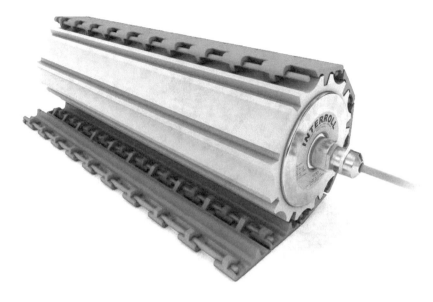

Fig. 2.9 Drum motor with profiled NBR lagging (Source: Interroll.com)

A sprocket wheel-driven modular belt may wear out a little faster since the forces exerted by the sprocket wheels always act only at the same, particular points.

A frequently used alternative to sprocket wheels are profiles made of 70°–80° Shore A soft NBR or PU. (See Fig. 2.9)

These profiles can drive modular belts with a pronounced side profile.

The softer, shock-absorbing rubber lagging drives the modular belt quietly. The continuous profile allows for power transmission across the entire width of the belt.

The force is thus optimally transmitted to the entire width of the modular belt, resulting in less mechanical strain on the belt.

From a hygienic point of view, however, modular belts are not quite as optimal as dirt, product residues and bacteria can collect between the hinges and modules. The many gaps and edges of a modular belt are usually not easy to clean. Occasionally undesirable bacteria may grow in the gaps and hinges that can not always be successfully combated even with chemical cleaning agents.

2.3   Hygienic, positive drive thermoplastic belts

The conveyor belts currently considered to be most suitable for maintaining hygienic conditions are positive drive thermoplastic belts. They are also often referred to as "blue belts". The colour blue is very commonly used in the food industry.

Blue is a colour that is very rare in natural foods.

Fig. 2.10 Drum motor and white thermoplastic conveyor belt (Source: Interroll.com)

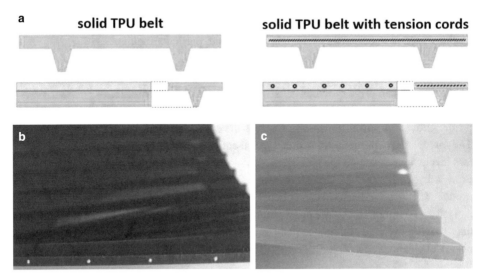

Fig. 2.11 a Structure of positive drive thermoplastic belts. **b** Belts with tension cords. **c** Belt with no tension cords

Fig. 2.12 Different spacings

If, for some reason, parts of the blue conveyor belt come off and get into the food product, then optical sensors can detect the blue foreign object relatively easily. The food production can then be stopped quickly if needed.

The positive drive thermoplastic belts are basically a combination of conventional flat belts and modular belts.

The upper surface is flat and closed. The underside is profiled for the positive power transmission. (See Fig. 2.10)

Thermoplastic belts are usually hot formed or cold milled in a single piece from TPU (thermoplastic polyurethane).

The TPU material softens when heated and returns to its original rigidity after it has cooled. The ends of a thermoplastic belt can easily be welded together to form an endless belt.

This results in a solid, smooth and completely closed conveyor belt, which consists of only one single piece of TPU.

Some belt manufacturers also build tension cords into their thermoplastic belts.

The tension cords ensure that the thermoplastic belt only undergoes minimal elongation when loaded. This allows softer and more flexible TPU materials to be used, allowing for smaller minimum deflections. (See Fig. 2.11a, b and c)

Positive drive thermoplastic belts with approximately 1″ or 2″ spacing are the most common type to be found on the market. (See Fig. 2.12)

Belts with 1″ spacing are usually slightly thinner and allow for smaller deflections.

However, positive drive thermoplastic belts are not suitable for knife edges or narrow transfers.

In recent years, various belt manufacturers have also started to offer thermoplastic belts with smaller spacing of about 0.5″.

When used in conjunction with head-driven conveyors with drum motors, spacings of less than 1″ generally make little sense, since drum motors, even with the smallest shell diameter do not normally fall below the minimum deflection diameter of 1″ belts.

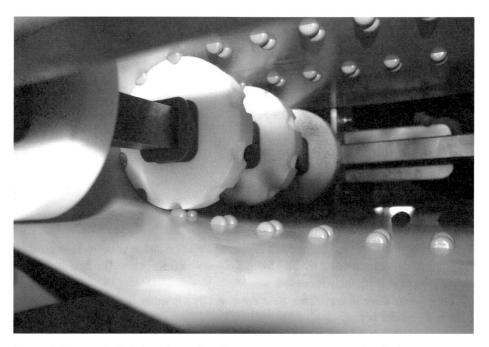

Fig. 2.13 Thermoplastic belt with round tooth contours (Source: Ammeraal Beltech)

Dirt and product residues can become trapped more easily in driver teeth that have a sharp tooth contour with sharp corners and edges.

Continuous driver teeth that extend below the belt across the entire belt width may distribute the force more evenly over the band, but they have a relatively large number of surfaces with corners and edges.

The rounder the tooth contour and the smaller the areas with corners and edges, the less dirt can be caught on the underside of the belt.

Positive drive thermoplastic belts can be operated without belt tension or with minimal belt tension.

This not only protects the ball bearings of the drum motor or drive drum, but an unstrained or very slightly tensioned belt can also be lifted easily for cleaning purposes.

This makes it possible to reach and clean areas that are usually difficult to access. (See Fig. 2.14)

Endless, welded thermoplastic belts typically have a smooth upper and lower surface with no gaps or cracks that could trap dirt and bacteria.

The belt can therefore be cleaned and disinfected quickly and easily.

Particularly in sensitive applications, such as the transport of raw and unpackaged fish, it is almost impossible to do without hygienic thermoplastic belts in modern applications.

But even the cleanest belt from a hygiene point of view makes no sense if it can not be hygienically driven.

Fig. 2.14 Cleaning a thermoplastic belt (Source: Intralox.com)

   Belt manufacturers usually supply the appropriate plastic sprocket wheels for positive drive thermoplastic belts.

These positively transfer the power of the drum motor or drive drum.

However, even with sprocket wheels there is always a small air gap between the drum shell and the wheel in which dirt can be caught, a particularly favourable environment for bacteria.

These bacteria are then rinsed out during the next cleaning process and could possibly contaminate the environment.

Sprocket wheels offer many advantages, but they are not the ultimate solution in terms of hygiene.

Solid, continuous and smooth profiles made of hygienic PU (82 Shore D) or stainless steel are a hygienic alternative to sprocket wheels. (See Fig. 2.15)

Fig. 2.15 Drum motor with premium Hygienic PU profile (Source: Interroll.com)

Solid Hygienic PU profiles have no sharp edges or gaps and are solidly cast or milled from a single material and should have a hygienically smooth surface roughness of less than Ra 0.8 μm.

The smooth surface is easy to clean, as it does not provide any adhesive surface for dirt and bacteria.

The friction between the positively driven thermoplastic belt and the profile must be as low as possible to ensure that the power transmission is effected only via the positive interlocking and not through friction. (See Fig. 2.16)

If the positive drive thermoplastic belt is driven by friction, then the belt speed increases slightly relative to the drum speed, because frictional engagement increases the unwind roll diameter of the belt slightly.

The larger unwind roll diameter allows the thermoplastic belt to move at a slightly faster speed compared to the driving profile.

When this happens, the thermoplastic belt eventually overtakes the profile until it collides with the teeth and, in the worst case, pops out of the profile.

In order to avoid undesirable frictional engagement between thermoplastic belts and the driving profiles or sprocket wheels, one should always observe and adhere to the current installation guidelines of the belt manufacturer.

Most belt manufacturers recommend that the belts are installed fully tension-free or with only minimal belt tension.

As a result, positive drive thermoplastic belts generally hang down further in the lower strand.

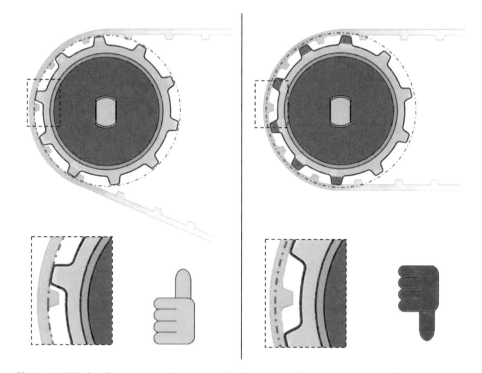

Fig. 2.16 Frictional engagement between TPU belt and profile should be avoided

This belt sagging is known as "catenary sag".

On account of this catenary sag, positive drive, thermoplastic belts are not suitable for tail drive conveyors.

Positive drive thermoplastic belts should always be pulled and if at all possible, never pushed. (See Fig. 2.17.a & b)

Definition of terms Fig. 2.17a and b:

CS: "Catenary sag" belt sagging

SR: Snub roller

Positive drive thermoplastic belts can usually be operated bidirectionally on conveyors with centre drive.

It is important to make sure that catenary sag can occur in both directions of rotation in the lower strand.

Occasionally, positive drive thermoplastic belts are closed to form endless belts using quick connectors. (See Fig. 2.18)

The quick connectors are similar to the hinge connection in a modular belt.

While not necessarily hygienic, these quick connectors make sense, for example, in conveyors found in narrow, hard to access machines.

All that is required to disassemble the belt is to pull out the hinge pin and the endless belt is opened and can be pulled out of the conveyor with ease.

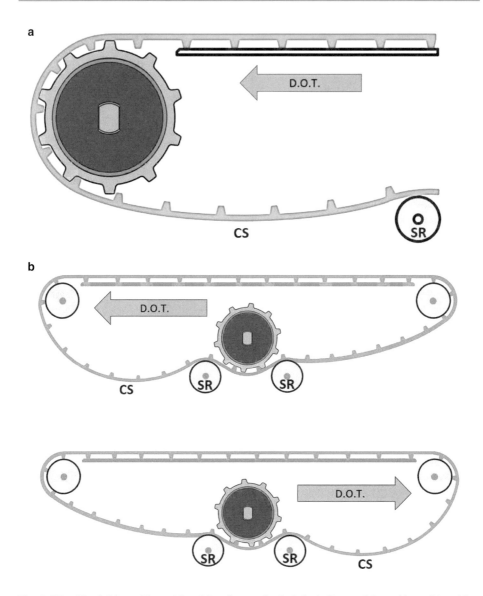

Fig. 2.17 a Head drive with positive drive thermoplastic belt. **b** Centre drive with positive drive thermoplastic belt

Having been removed, the belt and the supposedly unhygienic quick connection can then be thoroughly cleaned and disinfected.

For belts with quick connectors, it is important to check that they are compatible with the profile of the drum motor or drive drum.

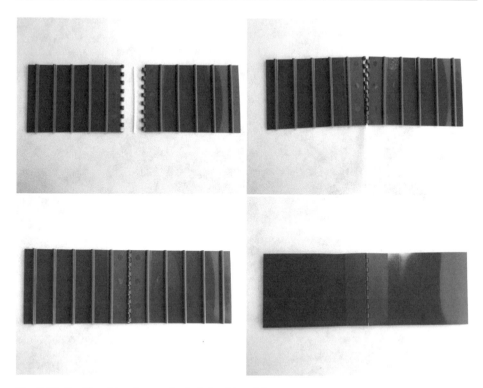

Fig. 2.18 Positive drive thermoplastic belt with quick connector

Common conveyor drives 3

In the preceding chapters you were introduced to various types and configurations of belts. The heart of a conveyor, however, is the drive. It brings the conveyor to life.

Three-phase asynchronous motors are one frequently used type of drive. They can easily be connected to a local three-phase network or to a variable frequency drive (VFD) to change their rotational speed.

Single-phase capacitor motors are also still used in light applications, such as conveyor belts at supermarket checkouts. These drives only make sense if no three-phase network is available and it is not possible to use a VFD. This technology is already obsolete for most industrial applications.

Since the prices for VFD electronics have become increasingly more affordable in recent years, more and more synchronous servomotors are emerging on the market, which now enable conveyor applications that seemed unthinkable in the past. Synchronous servomotors are extremely energy-efficient, dynamic and versatile. The future of conveyor technology lies with synchronous servomotors on account of their extremely short acceleration and deceleration times, very broad speed control ranges and high power densities.

Two types of motor currently dominate modern intralogistics: The drum motor, which can be integrated almost invisibly into a conveyor system and is therefore simply overlooked by many people, and the conspicuous shaft-mounted geared motors, which are mounted on the side of or beneath a conveyor belt assembly and drive a drive pulley.

© Springer-Verlag GmbH Germany, part of Springer Nature 2020
S. Hamacher, *The Drum Motor*,
https://doi.org/10.1007/978-3-662-59298-4_3

3.1   Drive pulley with shaft-mounted geared motor

Most people who work with conveyor belts will inevitably encounter geared motors that drive a drive pulley.

A shaft-mounted motor consists of two major components. (See Fig. 3.1)

① The electric motor – in many cases a three-phase asynchronous motor

② The gear mechanism – very commonly a worm drive or bevel gear

The electric motor is usually flange-mounted to the gear unit.

The rotor shaft of the electric motor rotates at a relatively high speed.

Depending on the number of pole pairs of the asynchronous motor, the speed of the rotor shaft is between approx. 450 min^{-1} (with 12-pole motors) and approx. 2800 min^{-1} (with 2-pole motors) (50 Hz grid).

However, these speeds are far too high for conveyor technology. This is why a gear mechanism is needed.

The gear unit has the task of reducing the high rotational speed of the rotor shaft while increasing the torque.

The shaft-mounted motor is often mounted on the side of or below the conveyor in a conveyor belt.

This results in the loss of valuable space because the shaft-mounted geared motor extends the dimensions of the conveyor laterally or vertically.

In order not to extend the conveyor frame even further, the rotational movement of the rotor is redirected by 90° in the gear unit in shaft-mounted geared motors.

This results in a great deal of frictional loss, particularly with worm gears.

Many worm gears therefore have an efficiency of only 50–60%.

This means that the power of the motor driving the worm gear has to be almost twice as high as the power required to act on the drive pulley.

This not only expends unnecessary energy but also often means that the electric motor has to be dimensioned one size larger.

The shaft-mounted geared motor on a conveyor belt drives a drive pulley.

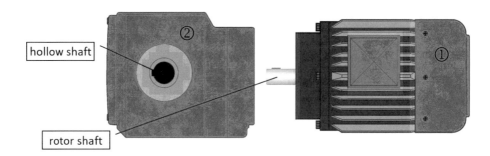

Fig. 3.1 Gear mechanism and electric motor

Fig. 3.2 Drive pulley with external ball bearings

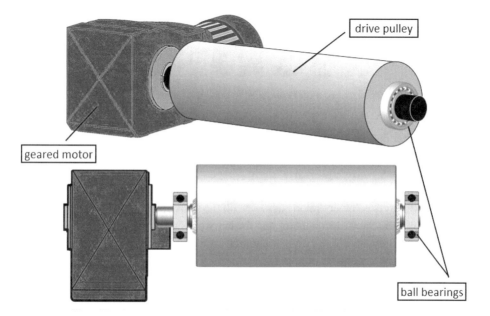

Fig. 3.3 Shaft-mounted geared motor with drive pulley and external ball bearings

The drive pulley is connected in many cases via an extended shaft with shaft key to a hollow shaft on the gear unit.

The drive pulley shaft is fixed to the drum shell, so the drive pulley must have its bearings mounted externally, since the entire axle must rotate with it. (See Fig. 3.2 and 3.3)

In wet and damp applications, such as are common in the food industry, the ball bearings must be regularly regreased after cleaning, since the ball bearing lubrication can be rinsed out by water.

This means additional maintenance work for the user. Regular lubrication means that lubricants are regularly discharged into the surroundings. The dispensing of lubricants in the food processing area is always problematic and undesirable in hygienic food applications. A frequently misinterpreted advantage of the shaft-mounted geared motor is the apparently rapid disassembly of the easily accessible motor. In practice, however, experience

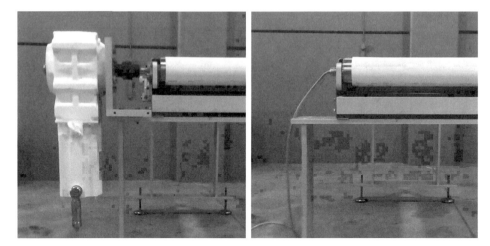

Fig. 3.4 Shaft-mounted geared motor compared to the drum motor (Source: Interroll.com)

has shown that the steel shafts of the drive pulley often rust inside the hollow shaft of the geared motor.

Even drive pulley with stainless steel shafts can get stuck in the hollow shaft of the geared motor if dirty water introduces small foreign particles into the fitting and blocks it.

This usually leaves no other option but to saw off the shaft of the drive pulley and to replace both the shaft-mounted geared motor and the drive pulley.

Shaft-mounted motors are certainly a working solution to drive a conveyor belt for dry logistics applications but the simple IP20 or IP54 shaft-mounted motors are not suitable for humid and wet applications.

However, now and then one does see fan-cooled geared motors in food applications that need to be cleaned with water on a daily basis.

If water were to get into the geared motor through the fan, it could cause a short circuit or even endanger people's lives. Geared motors are therefore preferably covered with elaborate stainless steel hoods in the food industry.

The stainless steel hoods must be partially open for the fan cooling of the motors, so time and again dirt and water from the occasional splash of water do get under the cover, an area that is seldom or never cleaned. Bacteria can then multiply there, which are then blown throughout the entire food production area by the geared motor's fan.

Modular belt conveyors used in food applications should run during cleaning, so that the hinges of the modular belt can be opened and cleaned.

IP20 or IP54 geared motors mounted on the conveyors without a stainless steel cover must then be provisionally waterproofed by the cleaning staff using a plastic bag.

If the fan sucks the plastic bag onto the ventilation grating, no more cooling air can flow into the geared motor. The geared motor overheats and burns.

Fig. 3.5 Rusty IP20 shaft-mounted motor under a stainless steel cover

The geared motors are simply painted, so rust spots often form in wet environments. (See Fig. 3.5)

Rusty parts should always be avoided in a food production area.

Simple IP20 or IP54 geared motors are therefore not really appropriate in the open food industry.

Enclosed IP66 or IP69k geared motors made of stainless steel are an alternative to fan-cooled geared motors. Their disadvantage, however, is that they can only be surface-cooled. This means they get very hot.

Since they are easily accessible, there is a risk that people could burn themselves on the hot motor surface.

The environment is often chilled in open food production areas. Ineffective, heat-generating geared motors are therefore counterproductive.

Over and above the losses that the motor and above all the gear mechanism already generate, even more energy must be expended for cooling the environment.

In short, the geared motor and drive pulley system is not the optimal solution for wet and hygienic conveyor belt applications in open food production areas.

3.2   Drum motors

As described in the previous section, shaft-mounted geared motors have several disadvantages, particularly in humid, wet and hygienic applications as well as in applications with limited space.

These disadvantages were analysed and the drum motor has emerged as a solution.

The drum motor combines the electric motor, the gearing and the bearing-mounted drive pulley in *one* compact, hermetically sealed component.

The drum motor can therefore be perfectly integrated into a conveyor belt to save space. The drum motor can hardly be seen from the outside, it is virtually invisible.

The encapsulated design of the drum motor has an inherently very high level of protection against the ingress of impurities and water. As a rule, drum motors have protection class IP66 or IP69k (see Fig. 3.6). They are therefore ideally suited to withstand regular cleaning processes with water and cleaning agents. Even cleaning with a high pressure water jet is no problem for a drum motor.

The first industrial-grade, mass-produced drum motors were developed in Denmark in the early 1950s by John Kirkegaard, who then established the company JoKi (The initial letters from **Jo**hn and **Ki**rkegaard).

JoKi became part of the Interroll Group in 1987.

Drum motor technology has steadily improved and been developed further since the 1950s. Today's drum motors are complex, state-of-the-art machines that are optimally adapted to the demands of modern industrial applications.

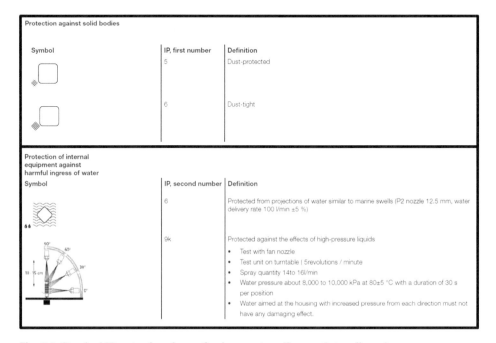

Fig. 3.6 Standard IP protection classes for drum motors (Source: Interroll.com)

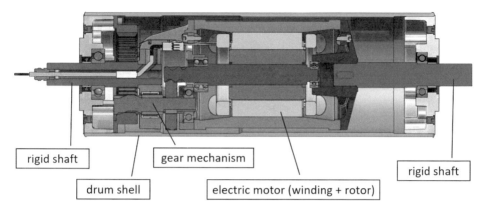

rigid shaft

gear mechanism

rigid shaft

drum shell

electric motor (winding + rotor)

Fig. 3.7 Structure of a drum motor

In classic drum motor construction the rigid shafts, the electric motor housing and the gear housing form a rigid, continuous axis.

The rigid shaft protrudes left and right out of the drum motor and usually has spanner flats on both sides. (see Fig. 3.7)

The shaft is placed with the spanner flats left and right in two brackets. The full torque of the motor acts on the brackets. The brackets must therefore be appropriately dimensioned to be stable enough for the forces acting on them.

It is therefore important that there is very little, or better still, no torsional backlash between the drum motor shaft and the brackets to prevent the drum motor shafts or brackets being knocked out.

The rigid drum motor shafts are thus often clamped into the brackets with screws to prevent play. (See Fig. 3.8)

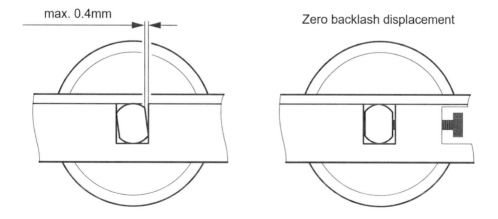

max. 0.4mm

Zero backlash displacement

Fig. 3.8 Drum motor mounting, torsional backlash

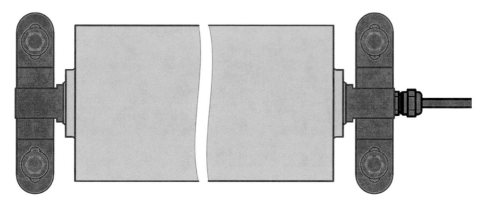

Fig. 3.9 Drum motor mounting from above

The brackets should not be made too thin. As a rule, the brackets should cover approximately 80% of the spanner flats. (See Fig. 3.9)

The rotor of the electric motor rotates at a relatively high speed. A drum speed of, for example, 2800 min^{-1} would be much too fast for an application in conveyor technology, however. This is why a gear mechanism is needed. The gearing converts the high rotor speed into a lower, usable speed, delivering a higher torque to the drum.

Highly efficient helical or planetary gears are preferred in drum motors.

Depending on the type of gear and the number of gear stages, the efficiencies of these gear units range from around 85–95%.

A drum motor can thus apply the same torque at the same speed with significantly lower energy consumption than would be possible, for example, with a shaft-mounted geared motor with worm gears. The efficiency of worm gears is only about 50–60%.

In the final gear stage, an output pinion drives a crown gear. The crown gear is firmly connected to the output cover. The output cover is in turn fixed to the drum shell. (see Fig. 3.10)

In this way, the force of the electric motor or the gearing is transmitted to the drum shell.

Drum motors are usually filled with oil in order to lubricate the mechanical parts such as gears and ball bearings in the drum motor and to better transfer the waste heat of the electric motor to the shell.

The squirrel-cage rotor is mounted on rotary bearings in the centre of the three copper coils of the asynchronous winding. This makes modern drum motors largely maintenance-free, as there is no need for regular relubrication.

In light, dry to damp applications, drum motors with seals that can be relubricated are occasionally still used since the sealing system on the shaft is more cost-effective to implement in this design.

Using regreasable seals in a drum motor, however, eliminates the drum motor's hygienic advantage and is not a maintenance-free option.

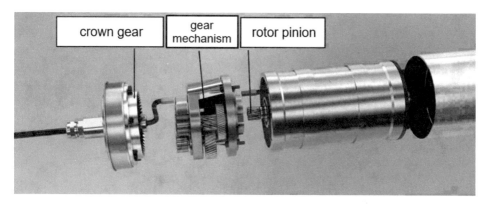

Fig. 3.10 Structure of a drum motor helical gear mechanism (Source: Interroll.com)

Modern drum motors are therefore no longer constructed with regreasable seals. (See Fig. 3.11)

Highly efficient drum motors with synchronous motor technology represent the latest development in drum motor technology.

Modern synchronous drum motors can already be made that are not filled with oil due to their high efficiency and thus lower heat generation. Mechanical parts such as gears and ball bearings are then enclosed and built with a grease filling.

Oil-free drum motors are sometimes used in hygienic applications where the drum motor could come in direct contact with open food.

The possibility of oil leaks can practically be excluded nowadays with well-known drum motor manufacturers and assuming proper use.

The sealing systems are tried and tested.

Unfortunately, this statement does not always apply to cheap imitation drum motors.

The quality of a drum motor is easy to gauge, among other things, by the quality and complexity of the design of the sealing system.

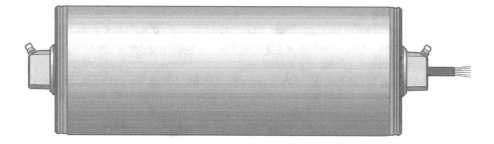

Fig. 3.11 Drum motor with grease nipple (Source: Interroll.com)

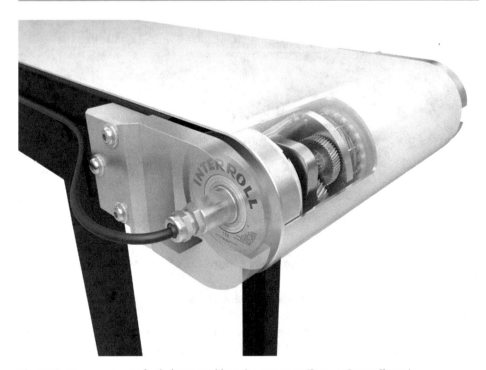

Fig. 3.12 Drum motor perfectly integrated into the conveyor (Source: Interroll.com)

Occasionally, however, the ball bearings of the drum motor may still become overloaded by excessive belt tension, which can result in oil leakage.

If metal parts come loose in the ball bearing due to overloading, they can damage the sealing system. The user then first notices an oil leak, without necessarily recognising the prior ball bearing damage due to belt tension.

The drum motor has therefore unfortunately acquired the unjustified reputation of not always being sealed. When handled correctly, however, this statement is absolutely false.

Drum motors have many different applications. The classic use of drum motors is as a belt drive in conveyors.

Since the drum motor can be completely integrated into the conveyor, it does not extend the construction laterally. (See Fig. 3.12) There is consequently no danger that someone can get burned on a hot, protruding motor.

Accidental damage to the belt drive by a careless forklift driver, for example, can also be excluded, because they would have to drive right into the conveyor to damage the drum motor.

Drum motors are thus much more reliable to operate in robust industrial applications than shaft-mounted motors.

3.3 Asynchronous drum motor technology

Asynchronous motors are by far the most common form of electric motors currently used in industrial applications. They are robust and can be manufactured comparatively cheaply. They are highly efficient and can be operated at constant transport speeds without additional control electronics. Asynchronous motors are therefore usually the first choice for applications which involve a steady flow of medium-weight goods within normal speed ranges.

Fundamental principles

To understand the complex principle of the asynchronous motor, you first have to know some basics:

1. *Definition of terms:*

   V = Volts, electrical voltage

   A = Ampere, electric current strength

   AC = alternating current/voltage

   DC = direct current/voltage

2. *Voltage:*

   Voltage (potential difference) exists wherever there is a difference in electrical charge or potential between two points.

   The voltage represents the tension between the differently charged points as they attempt to balance each other out.

   The greater the difference in charge or potential between the two points, the greater the voltage.

3. *Current:*

   Electric current is the directed movement of free-moving charge carriers.

   If, as in point 2 (voltage), points with different charges are connected to each other with an electrical conductor such as a copper wire, the excess charge of the point with higher potential flows through the copper wire to the point with lower potential until both points are equally charged. The flow of charge carriers is referred to as electric current.

   The number of charge carriers that can flow through the copper wire in a given time, depends on the resistance of the copper wire and the voltage.

   The higher the number of charge carriers that are flowing, the higher the electric current.

4. *Electromagnetism:*

   A current-carrying electrical conductor produces a magnetic field around itself.

   When an electrical conductor is wound, the magnetic field is amplified when current flows through it.

5. *Electromagnetic induction:*

   If a magnetic field brushes against an electrical conductor, a voltage is generated in the conductor at the moment the magnetic field changes. This voltage is also referred to as the induced voltage.

   In order to continuously generate an induced voltage in the conductor, the magnetic field must remain in constant motion.

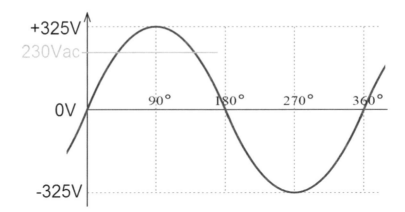

Fig. 3.13 Example of an alternating voltage cycle

6. *Alternating voltage:*

   Alternating voltage is a continuously changing voltage, which can be represented in the form of a sine curve.

   The curve starts at 0°, the voltage being 0 V. The voltage increases sinusoidally until it reaches its maximum positive voltage value at 90° (e.g. +325 V). After the maximum voltage has been reached, the voltage drops sinusoidally until it reaches 0 V again at 180°. At 180°, the voltage is reversed.

   The voltage continues to decrease sinusoidally until it reaches the maximum negative voltage value at 270° (e.g. −325 V).

   From 270° the voltage increases again sinusoidally until it reaches 0 V again at 360°. At this point, the voltage is reversed again and the process begins once more.

   A sinusoidal alternating voltage with +325 and −325 V as peak values is referred to as 230 Vac alternating voltage.

   230 Vac corresponds to the average effective voltage, which results from the alternating voltage of +325 and −325 V. (See Fig. 3.13)

7. *Frequency:*

   The frequency of an alternating voltage indicates the speed at which the alternating voltage changes or oscillates.

   The unit of frequency is Hertz [Hz]. 1 Hz = 1 sine curve per second.

8. *Three-phase alternating voltage:*

   Three-phase alternating voltage is a voltage consisting of three individual alternating voltages of the same frequency.

   The three individual alternating voltages are phase-shifted by 120° relative to one another.

   The three-phase alternating voltage can generate a so-called rotating field with little effort.

   A rotating field is necessary for an asynchronous motor to rotate. (See Fig. 3.14)

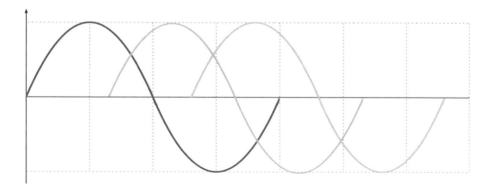

Fig. 3.14 Example of a three-phase alternating voltage

Structure of the asynchronous motor

The asynchronous motor consists of two main components.
1. The stator, also called the winding.
2. The rotor, also known as a cage or squirrel-cage rotor.

The stator

A copper wire insulated with lacquer is wound into a coil. (See Fig. 3.15)

Depending on the design, a coil such as this can have several hundred turns.

A copper coil amplifies the electromagnetic field of a current-carrying conductor. This means that a coiled wire generates a significantly higher electromagnetic field than an unwound wire of the same length.

If an alternating voltage is connected to the copper coil, an electromagnetic field is initially created by the flow of current in the copper wire.

The constantly changing value of the alternating voltage results in continuous change in the strength and orientation of the magnetic field. (See Fig. 3.16)

Abb. 3.15 Copper coil for the generation of electromagnetic fields

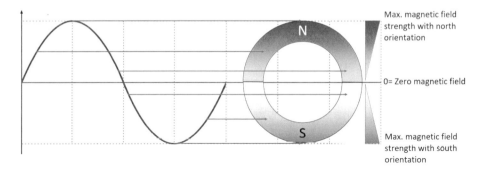

Fig. 3.16 Magnetic field of a copper coil connected to alternating voltage

In an asynchronous motor winding, three copper coils are spatially arranged at an angle of 120° to one another.

The three coils are marked with the letters U, V and W.

The beginning of a coil is labelled 1 and the end 2.

One phase of the three-phase alternating voltage is connected to each of the three copper coils.

The alternating voltages phase-shifted by 120° now also act in the 120° spatially offset coils. (See Fig. 3.17)

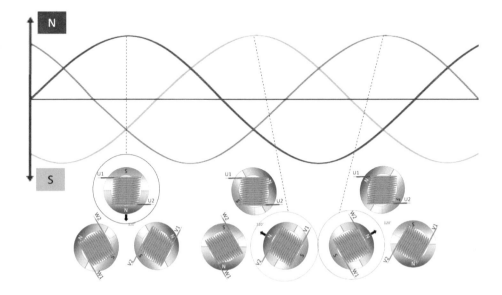

Fig. 3.17 Electromagnetic field generation by a three-phase alternating voltage

Fig. 3.18 Classic asynchronous winding in the iron packet

This results in time-staggered, constantly changing electromagnetic fields with constantly reversing polarity that are coordinated in such a way that a rotating electromagnetic field is produced between the three copper coils.

The copper coils are wound around iron packets to strengthen and direct the magnetic field lines.

These iron packets found in modern asynchronous windings, consist of several thin sheet metal layers which minimise the loss-inducing, heat-generating eddy currents or so-called iron losses arising from undesirable electromagnetic induction in the iron core. The sheet metal layers are electrically insulated from each other and are joined with rivets or spot welds to form a solid iron packet. (See Fig. 3.18)

The Rotor

The craziest component on the asynchronous motor is definitely the rotor, also known as the cage or squirrel cage rotor.

A so-called squirrel cage is, as the name suggests, built like a round cage. (See Fig. 3.19) The squirrel cage, like the asynchronous winding, consists of a laminated iron core. Several grooves are punched out in the iron core, which are usually filled with cast aluminium. An aluminium cage is thus constructed around the insulated iron core.

The cage bars are electrically shorted at both ends by closed aluminium rings.

The squirrel cage rotor is sometimes also called a "short circuit rotor" because of the bars being short circuited. (See Fig. 3.20)

The squirrel cage rotor is mounted on rotating bearings in the centre of the three copper coils of the asynchronous winding. (See Fig. 3.21)

Fig. 3.19 Basic structure of
a "squirrel cage" rotor

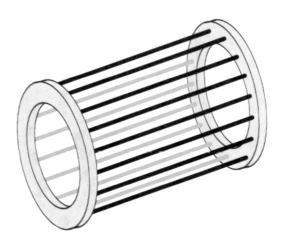

Fig. 3.20 An actual squirrel
cage rotor

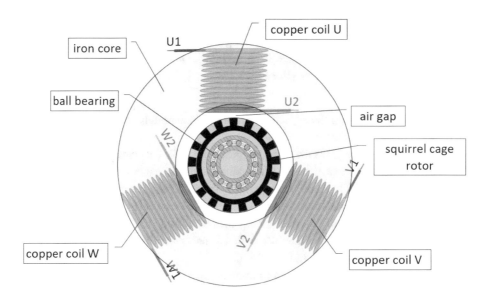

Fig. 3.21 Basic structure of an asynchronous motor with squirrel cage rotor

Asynchronous motor operating principles

Since the electromagnetic fields of the three copper coils are changing constantly, an induced voltage is generated in the bars of the rotor's aluminium cage according to the principle of electromagnetic induction.

The amount of voltage induced depends on the intensity of the electromagnetic field around each cage bar and therefore differs from bar to bar.

As a result, the induced voltages of the individual bars have different levels of voltage potential.

The potential difference between the bars is balanced out by the short circuit at the ends of the bars.

The equalisation of potential results in a directional movement of charged particles, which ultimately means that an electric current flows through the cage bars.

According to the principle of electromagnetism, a current-carrying conductor generates an electromagnetic field.

This then means that individual magnetic fields form around the bars of the squirrel cage, which are in turn attracted to the rotating magnetic fields of the asynchronous winding.

It is this magnetic attraction that ultimately causes the squirrel cage to move. (See Fig. 3.22)

Building up the electromagnetic field in the squirrel cage rotor takes a certain amount of time.

The time-staggering between the electromagnetic field of the copper coils on the stator and the electromagnetic field generated around the rotor by induction means that the squirrel cage rotor always rotates somewhat slower than the surrounding, revolving magnetic field in the stator.

The rotor can never synchronise with the revolving magnetic field of the stator. Its rotational speed is therefore asynchronous relative to the stator's magnetic field.

This is why this design of motor is called an asynchronous motor.

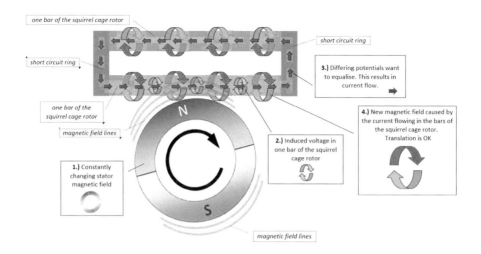

Fig. 3.22 Electromagnetic induction

The difference between the rotational speed of the stator's magnetic field and the rotor speed is called slip.

If the slip increases, for example by increased mechanical load on the rotor shaft, then the current in the copper coils in the stator automatically increases, which in turn produces a stronger electromagnetic field in the stator and in the rotor.

The stronger magnetic fields in the stator and rotor in turn result in an increase in the amount of torque that can be mechanically retrieved at the rotor shaft.

The increased current and the resulting stronger magnetic force generate greater heat losses, so the mechanical load on the rotor shaft may not exceed a certain point to avoid the asynchronous motor being permanently overloaded.

This point is called the operating point. The resulting current in the copper coils of the stator at this point is called the rated current.

The operating point is generally the point at which the asynchronous motor works most efficiently with respect to the applied and output power.

The big advantage of the asynchronous motor is its simple and robust design and that the motor can start independently on a three-phase alternating voltage grid.

The disadvantage, however, is that they operate relatively inefficiently, which is particularly true for smaller asynchronous motors.

Short circuiting the induced voltage in the rotor generates losses that are converted into heat.

The efficiency of an asynchronous motor is defined, among other things, by the ratio of the air gap between the rotor and the stator to the mass of the rotor.

Since the air gap cannot be made infinitely small for technical, manufacturing reasons, the air gap in small asynchronous motors is much larger in relation to the mass of the rotor than in larger, higher power motors.

The larger the air gap in relation to the rotor mass, the more waste heat the asynchronous motor generates.

Disadvantage of the squirrel cage rotor

The laminated, insulated iron core in the rotor, which is surrounded by the aluminium cage, is needed to amplify the electromagnetic field of the rotor and to guide it in the right direction.

The weight of the iron core, however, makes the rotor heavy, which gives the rotor a high level of inertia.

Accelerating an object of great inertia is always associated with a slow start-up. This means that, especially in highly dynamic applications where asynchronous motors are starting up, a great deal of energy is always required if the rotor is to be accelerated quickly.

Since the rotor is not in motion at the instant of switching on, the slip is momentarily 100%. This means that for a short time 3–8 times the rated current must flow through the stator windings to generate an overexcited magnetic field that can overcome the moment of inertia of the rotor and the load connected to it.

If a lot of energy is needed, this automatically means that more waste heat is also generated. In the long term, an application that is too dynamic, in which an asynchronous motor has to be started and stopped quickly and frequently, will lead to severe heating and even overheating of the motor.

Asynchronous motor technology thus has its limits, especially in dynamic applications.

Types of asynchronous motors

Asynchronous motors can be made as 2-, 4-, 6-, 8- or 12-pole motors.

In each variant, the copper coils are wound or arranged differently in the stator.

2-pole windings have 1 pole pair, 4-pole windings 2 pole pairs, 6-pole windings 3 pole pairs etc.

The rotor speed depends on the frequency of the applied three-phase alternating voltage and on the number of pole pairs on the stator. (See Fig. 3.23)

The rotor speed of an asynchronous motor is calculated as follows:

ns =   synchronous speed at the stator

s =   slip

f =   frequency of the voltage

p =   number of pole pairs

nr =   rotor speed

Formula:

$$nr = \frac{f \times 60s}{p} - s$$

Stator windings for different numbers of poles allow the rotor speed to be roughly adapted to the respective application.

The fewer poles the winding has, the more efficiently it operates.

2-pole windings are usually the most efficient.

Nevertheless, 4-pole asynchronous motors are often preferred in conveyor technology, as they are characterised by their very smooth operation and they work efficiently enough.

6-, 8- and 12-pole windings are usually only used where very slow speeds with a lot of torque are required. One does need to pay attention to the cooling of these motors when

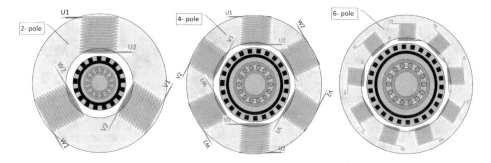

Fig. 3.23 Structure of asynchronous stators with different numbers of poles

dimensioning a system, however. When it comes to very small asynchronous motors (under 100 W), 6-, 8- and 12-pole asynchronous motors are extremely inefficient and produce more heat than mechanical power. As a result, 6-, 8-, and 12-pole motors are almost never found in the low power range.

As a rule, the more poles an asynchronous motor has, the more expensive the stator winding.

This is partly because significantly more of the expensive copper has to be used for multi-pole windings. In addition, multi-pole windings take considerably more effort to manu-facture.

If one considers that 18 coils must be accommodated in the stator for a 12-pole winding and only 3 coils in a comparable installation space for a 2-pole winding, it is easy to see why multi-pole windings are more expensive in comparison.

Asynchronous motor cooling

As one can never completely avoid losses that are converted into heat with asynchronous motors, it is important to provide adequate cooling for the motor.

In conventional geared motors, a fan wheel is therefore mounted on the rotor shaft and the stator housing is made permeable to air, so that cooling air can be drawn in at one end and blown out at the other end. (See Fig. 3.24)

Air cooling works for standard applications but has some drawbacks in applications where the rotary speed has to be controlled by variable frequency drives.

If an air-cooled geared motor with a variable frequency drive is operated very slowly, the circulation of air may not be sufficient, because the fan wheel is not rotating fast enough.

If a fan-cooled asynchronous motor is operated above its rated speed, the more quickly rotating fan wheel generates a larger air resistance, which slows down the rotor. The greater air resistance of the rapidly rotating fan therefore decreases the efficiency of the motor.

Dust, dirt and moisture can also be sucked in through the louvres in the stator housing.

A fan-cooled motor should especially not be used in wet or dusty environments.

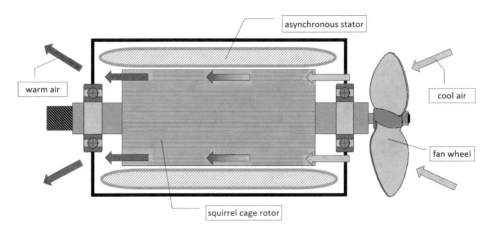

Fig. 3.24 Asynchronous motor with fan cooling

Fig. 3.25 Asynchronous drum motor with belt cooling

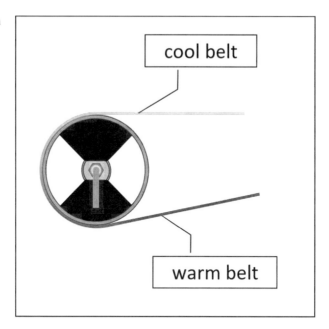

The cooling of a drum motor, on the other hand, works in a completely different way.
The drum motor is completely hermetically sealed and has a high protection class of IP66 or IP69k, so no dust or moisture can penetrate the drum motor.
Fan cooling therefore makes no sense in an enclosed space.
A drum motor has oil cooling, which fulfils three tasks simultaneously.
1. The oil lubricates mechanical components such as gears or ball bearings.
2. The oil carries the heat from the winding to the drum shell.
3. The oil provides for a more uniform heating of the drum shell.

Belt drives are the classic application for drum motors.
The conveyor belt can be used to cool the motor. During operation, it absorbs some of the drum shell's heat. (See Fig. 3.25)
The conveyor belt can cool down in the lower and upper strands, until it is cool enough after one revolution to absorb heat from the drum shell again.
The belt thus constantly draws the heat away from the drum shell.
This principle works even if the drum shell has been coated with rubber lagging of up to 8 mm thick.
Depending on the motor variant, load and ambient temperature, the drum temperature of an asynchronous drum motor, with incorporated belt cooling, can be approx. 30–60° C.
Drum motors can also be used in applications without belt cooling.
The drum motor just has to be dimensioned more generously for this.
Although the asynchronous motor is no longer operated at operating point when it is over-sized, it also generates less waste heat owing to the lower power output. In other words, the motor operates at a lower temperature overall.

The heat of the drum shell can now only be released to the surrounding air.

An application using a drum motor without a belt cooling system should therefore not operate in an environment that is warmer than +25° C.

Depending on the motor variant, load and ambient temperature, the drum temperature of an asynchronous drum motor without belt cooling and with approx. 10–20% power reserves, can be approx. 70–90° C.

As a rule of thumb, as long as the motor's rated current is not exceeded and as long as the winding protection contact does not trip, everything is okay.

3.4   Synchronous drum motor technology

Although the stator of a synchronous motor is usually constructed somewhat differently to an asynchronous stator, the operating principle of the two stators is similar.

The synchronous stator essentially consists of copper coils and laminated iron cores.

The main difference in construction between asynchronous and synchronous motors lies in the rotor.

Very strong neodymium permanent magnets are embedded within a laminated iron core or glued on the outside of the rotor of a synchronous drum motor. As a result, the rotor of the synchronous motor has its own permanent magnetic field.

The need to elaborately build up a magnetic field in the rotor at great expense of energy, like in the squirrel cage rotor of an asynchronous motor, is thus completely eliminated.

This is one of the main reasons why synchronous motors can operate much more efficiently than asynchronous motors.

The squirrel cage rotor of an asynchronous motor has a large mass and surface area in order to generate a sufficiently strong magnetic field in the rotor.

Since the synchronous rotor already has a strong magnetic field, it can be made significantly smaller and lighter. (See Fig. 3.26)

Synchronous motor operating principles

A three-phase alternating voltage generates a rotating magnetic field in the synchronous stator similarly to how this is done in an asynchronous motor.

The synchronous rotor has its own, permanent magnetic field thanks to the neodymium permanent magnets mounted in the rotor.

The permanent magnetic field is attracted by the surrounding, rotating magnetic field of the synchronous stator.

Since no time is required to generate the magnetic field in the synchronous rotor, the rotor can synchronise with the magnetic field of the stator.

The rotor thus rotates at the same speed as the stator's surrounding, revolving magnetic field.

That is why this motor technology is known as a synchronous motor.

   But the operating principles of the synchronous motor are not as simple as they sound at first.

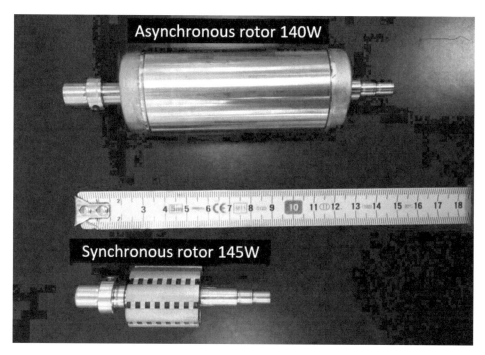

Fig. 3.26 Permanent magnet synchronous rotor compared to an asynchronous rotor

In order for the permanent magnet rotor to be able to synchronise with the revolving magnetic field of the stator, the synchronous motor must be ramped up as it starts.

If a synchronous motor is powered directly from the grid, a magnetic field is immediately created in the stator, that cycles at mains frequency. This is so fast that the permanent magnet rotor fails to synchronise with the magnetic field on account of its inertia.

If the synchronous motor is operated asynchronously to the stator magnetic field, it generates tremendous heat and the synchronous motor runs the risk of overheating or burning.
A synchronous motor must therefore never be operated directly connected to the mains.

It is imperative that a synchronous motor is always operated with a suitable variable frequency drive.

The frequency and voltage can be ramped up or down with the variable frequency drive.

However, a start-up ramp does not mean that the synchronous motor has to start up slowly. On the contrary, the ramp can be very short (0.1 s or shorter).

Although an asynchronous motor can be operated directly connected to the grid, its behaviour during start-up is quite sluggish. Thus, starting up an asynchronous motor, depending on its moment of inertia, is usually slower than starting a synchronous motor via a variable frequency drive with the smallest possible start-up ramp.

But the synchronous motor does not need a suitable variable frequency drive only for start-up only. Even during operation, the synchronous motor must be constantly monitored and readjusted as required.

For this purpose, the VFD constantly requires information about the current rotor position of the synchronous motor.

The VFD can use the current rotor position data to detect whether the rotor is rotating synchronously with the stator's magnetic field.

If the rotor speed changes due to, for example, load changes, the VFD detects this and can, if need be, influence the magnetising current by changing the voltage in the copper coils. When high torque is required, the VFD increases the voltage in the copper coils of the stator, resulting in a higher magnetising current and thus a stronger stator magnetic field. The synchronous rotor, which was previously decelerated by the higher load, is now attracted more powerfully by the stronger stator magnetic field until the synchronous speed is reached again.

The VFD thus continuously regulates the strength of the stator magnetic field, which needs to be constantly adjusted depending on the load.

Instantaneous rotor position data can be collected or the required speed control achieved in two main ways:
1. Closed control loop with encoder feedback.
2. Sensorless vector control.

The operating principles of these two types of control are explained in detail in the chapter "Variable frequency drives and encoder systems".

For a long time synchronous motors did not have a major role to play because there were no suitable drive options.

However, this has changed dramatically in recent years and there is now a very broad and cost-effective range of variable frequency drives and controllers on the market which can drive synchronous motors properly. The synchronous motor is thus becoming increasingly important in a wide range of industrial sectors.

Advantages of the synchronous motor
1. Synchronous motors are much more efficient because no energy needs to be wasted in generating a magnetic field in the rotor.
2. The smaller rotor makes the synchronous motor structurally smaller.
3. Since the synchronous rotor is smaller and lighter, it can be accelerated and decelerated more rapidly.
4. The motor speed remains virtually constant regardless of the load, owing to the speed control via variable frequency drive or servo drive.

As VFDs are already required for the majority of applications in conveyor technology nowadays, it is not really a disadvantage that one always has to use a VFD with a synchronous motor.

Synchronous motors have yet another specific characteristic that can be electrifying. Many people are not aware that a synchronous motor is also a generator at the same time. Thanks to the permanent magnet rotor, a mechanically driven synchronous motor generates voltage straight off without the need for additional tools or components.

Caution should therefore be exercised, especially during the installation of a synchronous drum motor in a conveyor belt, because the voltage that is produced on the motor's stranded wires, for example by manually moving the synchronous drum motor through the conveyor belt, can lead to an electric shock if touched.

The motor's stranded wires should therefore always be electrically insulated during installation of a synchronous motor.

Another much misinterpreted phenomenon that occurs with synchronous motors is when two or three stranded wires are shorted together.

A synchronous drum motor can only be rotated very sluggishly with short-circuited motor wires. One might assume that the problem is a mechanical defect in the motor.

However, once the short circuited motor wires have been dealt with, the motor can be rotated freely again.

The reason for this behaviour is the induced voltage generated by the synchronous motor during generative operation.

If the generator voltage is short-circuited on the motor's stranded wires, an opposing magnetic field develops in the motor winding, which slows down the motor.

Variable frequency drives and encoder systems

4

Variable frequency drives (VFDs) and rotary encoders have become indispensable in modern conveyor technology.

They are both complex electronic components, however, so that certain rules need to be adhered to.

Variable frequency drives can be differentiated into V/f controlled and sensorless control variable frequency drives.

Encoder systems are sensors mounted on the rotor shaft which can record information about the instantaneous rotor position and speed.

If a motor is used together with a rotary encoder and variable frequency drive in a closed control loop, it is usually referred to as a servo system or the VFD is then referred to as a servo controller.

The differences between the various types of variable frequency drives and encoders are explained in the following chapter.

4.1 Variable frequency drives

A variable frequency drive is required if the speed of a three-phase asynchronous or synchronous motor needs to be changeable.

There are variable frequency drives with different functions for a range of motor technologies and applications.

Basically, a variable frequency drive can be explained simply in terms of four main components: the rectifier, DC link, control circuit and inverter (See Fig. 4.1.)

© Springer-Verlag GmbH Germany, part of Springer Nature 2020
S. Hamacher, *The Drum Motor*,
https://doi.org/10.1007/978-3-662-59298-4_4

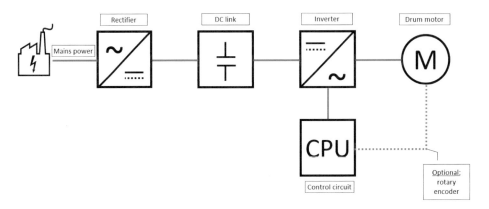

Fig. 4.1 Simplified block diagram of a variable frequency drive

The rectifier

The rectifier converts the AC voltage from the mains power to a pulsating DC voltage. (See Fig. 4.2.)

The rectifier converts the negative half wave of the voltage sine curve into a positive half wave.

This principle works with single-phase or three-phase AC voltage.

The variable frequency drive usually operates more efficiently if it is supplied with a three-phase AC voltage.

The DC link

The DC link consists mainly of capacitors. A capacitor can store electric charge, so its function is to smooth the pulsating DC voltage. (See Fig. 4.3.)

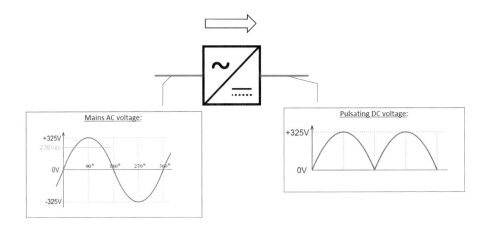

Fig. 4.2 Rectifier

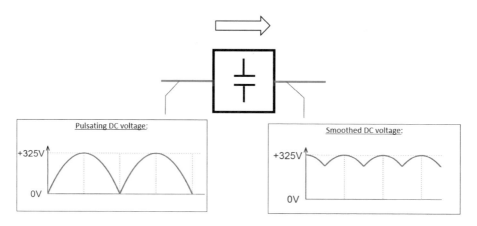

Fig. 4.3 DC link

The capacitor smoothes the pulsating DC voltage by releasing its stored electric charge whenever the pulsating DC voltage is in a trough.

If the pulsating DC voltage is at a peak, the capacitor recharges.

Having been smoothed by the capacitor, the DC voltage now only pulsates minimally.

A near perfect DC voltage can be modulated through further electronic measures.

The DC voltage measurement corresponds approximately to the peak value of the input AC voltage. (Factor $\sqrt{2} \times$ Vac)

For variable frequency drives with 230Vac supply, this results in a DC link voltage of approx. 325 Vdc.

For VFDs with a 400Vac supply, the DC link voltage is approx. 560 Vdc.

The inverter

The inverter is the heart of the variable frequency drive.

In the inverter, an AC voltage is modulated from the intermediate circuit DC voltage again. This type of AC production is called pulse width modulation.

As the name suggests, the DC link voltage is pulsed, i.e. only small voltage blocks are switched through. The width and polarity of the voltage blocks are controlled in such a way that a variable, rectangular alternating voltage is produced. (See Fig. 4.4.)

The inverter essentially consists of several high power transistors, so-called IGBTs. Transistors are electronic components that can switch through a voltage or a current contactlessly at a very high switching frequency.

Mechanical switching devices would not be suitable for the extremely high frequency switching cycles and would not withstand the stresses of rapid switching.

The IGBTs switch the DC link voltage at a speed of 2–16 kHz (2000–16,000 times per second). The switching frequency of the IGBTs is also known as the pulse frequency.

This variable, electronically generated voltage is connected to the copper windings in an asynchronous or synchronous stator where it results in a practically sinusoidal current flow.

The higher the pulse frequency, the smoother the sine curve of the motor current.

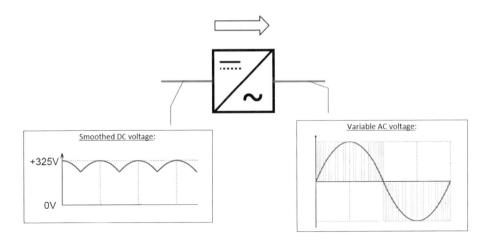

Fig. 4.4 Inverter

However, a high pulse frequency also causes greater losses in the variable frequency drive. A pulse frequency of about 8 kHz is recommended.

EMC compliant installation:
The high frequency switching of the IGBTs in the inverter does have a disadvantage, however. Every time an IGBT switches on the DC link voltage, a brief overvoltage peak occurs. These over voltage spikes cause something that can be described as high-energy electromagnetic radio waves that are emitted into the environment and can cause massive interference in other electronic components.

A system in which a variable frequency drive is installed must therefore be wired with EMC compliance. (EMC = Electromagnetic compatibility)
EMC means that care must be taken to ensure that the surrounding electrical and electronic equipment are not interfered with.

The electromagnetic radio waves generated by the VFD are sent into the environment through the motor cable. The motor cable thus acts like an antenna.
The longer the cable, the more serious the interference effect.
In order to prevent emission of the electromagnetic radio waves, the motor cable must be shielded.
The shielding is a closely-knit wire mesh, which must be earthed on both sides.
The high energy electromagnetic waves are taken up by the shield.
The resulting induced voltage is dissipated into the earth.
Whoever has touched an ungrounded shield during operation knows how much energy can be dissipated into the ground because touching an ungrounded shield can cause a serious electric shock.
The motor cable must therefore always be shielded and grounded on both sides.
As a rule, the motor cable for drum motors is already grounded inside the motor, so that the user only has to earth the cable shield at the connection side.

Rotary encoder cables should also always be equipped with a shield. The shield is used to protect encoders against electromagnetic interference that can come from the outside and affect the encoder. The screen of the encoder cable should be earthed at the same potential as the shield of the motor cable.

In addition to the high energy electromagnetic radio waves, interference can also spread via the supply line of the variable frequency drive into the power grid.
This so-called mains feedback can be counteracted with mains filters, which are connected upstream of the VFD input.

However, the mains filter usually causes a higher leakage current to earth, which can trigger standard commercial residual current device (RCD).
If it is not possible, for safety reasons, to do without a RCD in an application with a variable frequency drive, then a so-called "universal current sensitive RCD" should be used.
This unfortunately has the disadvantage that it is significantly more expensive compared to a standard RCD.

The control circuit

The control circuit can be described as the brain of the VFD. The control circuit consists of microprocessors which, among other things, control the switching of the IGBTs in the inverter.
Depending on the version of variable frequency drive, further complex calculations and control sequences are carried out in the control circuit.
The quality and performance of a variable frequency drive is very much dependent on the control circuit.

General principle for variable frequency drives

As already described in the preceding chapters, the motor speed is dependent on the velocity of the rotating magnetic field in the stator coils.
The velocity of the revolving magnetic field is defined by the frequency of the three-phase alternating voltage.
In other words, to change the speed of an asynchronous or synchronous motor, the frequency must be changed.
However, one must be sure to change the voltage too, while keeping it in the same ratio to the frequency.

Example:

Rated voltage and frequency of the motor:	$V = 400$ V; $f = 50$ Hz
Rotor speed at 50 Hz:	2750 rpm
Ratio V/f:	400 V/50 Hz = 8 V/Hz
VFD output frequency:	$f = 25$ Hz
VFD output voltage:	25 Hz $\times$ 8V/Hz = 200 V
Rotor speed at 25 Hz:	1375 rpm

In the example above, a 400 V, 50 Hz motor is operated at 25 Hz while connected to the variable frequency drive.

Fig. 4.5 Characteristic   V/f
curve

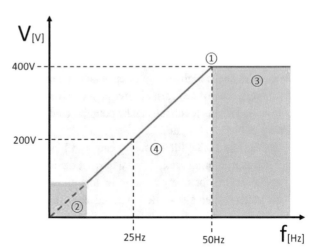

The motor speed and motor voltage change in proportion to the frequency.

Thus it must always be ensured that the ratio between voltage and frequency (V/f) corresponds to the ratio of rated voltage to rated frequency, so that the motor is not operated with overvoltage or undervoltage. (See Fig. 4.5.)

If an asynchronous or synchronous motor is operated with overvoltage, the current in the stator coils also rises and the motor can overheat or burn.

If an asynchronous or synchronous motor is operated with undervoltage, then the current in the stator and thus the electromagnetic force of the coils is lower.

The motor thus becomes weaker.

The overall torque behaviour of the motor changes. If the voltage is too low, the torque behaviour of the motor becomes unstable and it eventually fails altogether.

Explanation of Fig. 4.5.
① Vertex/motor rated frequency
② Lower frequency range
③ Field weakening region
④ Constant moment range

The characteristic V/f curve is linear. The slope of the V/f curve is defined by the vertex. The vertex can usually be determined from the rated motor voltage and rated frequency. The motor speed can then be regulated down from the vertex point.

But there are limits. Asynchronous motors in particular cannot be regulated down arbitrarily. Depending on the number of motor poles, the variable frequency drive's mode of control and the motor's power reserves, an asynchronous motor should not be regulated down too low.

If an asynchronous or synchronous motor is operated above the vertex, it runs in the field weakening region.

At the same time, the vertex is always the highest point of the V/f curve since it is always related to the maximum output voltage of the variable frequency drive.

From this point on, the voltage cannot be increased any further.

   Thus if a motor is to be run with frequencies higher than the vertex point, then the ratio of voltage and frequency will no longer be correct.

Example:

Rated voltage and frequency of the motor:	V = 400 V; f = 50 Hz
Rotor speed at 50 Hz:	2750 rpm
Ratio V/f:	400 V/50 Hz = 8 V/Hz
VFD output frequency:	f = 75 Hz
Rotor speed at 75 Hz:	4125 rpm
VFD output voltage/ratio:	400 V/75 Hz = 5.33 V/Hz

The ratio of voltage to frequency in the example is too small when the frequency exceeds 50 Hz. This means that the motor is operated with undervoltage above the vertex. The torque behaviour of the motor changes as a result. The motor has less torque.

If the motor is operated too far up in the field weakening region, the torque behaviour can become unstable until it eventually fails altogether.

Although the motor has less force in the field weakening region, it rotates faster due to the higher frequency.

The mechanical power of a motor is calculated from the rotational speed and the torque. If the force [F] of the motor decreases and the speed [v] increases in the correct ratio, then in the end the mechanical performance [Pmech] remains the same.

Formula:

$$P_{mech} = F * v$$

4.1.1 V/f controlled variable frequency drive

V/f controlled variable frequency drives are simple and usually inexpensive to purchase. They should only be used with asynchronous motors.

With V/f control, the output voltage is simply output statically corresponding to the set output frequency.

There is no communication between the motor and the variable frequency drive. The variable frequency drive does not respond to load changes on the motor in V/f control.

V/f control has no automatic control to influence the motor behaviour when changes occur. The behaviour of an asynchronous motor connected to a V/f controlled variable frequency drive is comparable to the behaviour of the motor when directly connected to the grid. (See Fig. 4.6.)

Since V/f control is very simple, there are not as many parameters required.

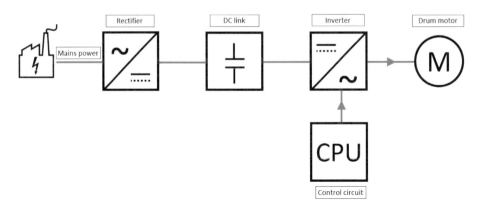

Fig. 4.6 Block diagram, variable frequency drive with V/f control

The most important parameters that must be set for V/f control are:
- the motor's rated voltage and rated frequency to determine the vertex.
- the *motor's rated current* to set the motor protection correctly.

Low output frequencies below 15–20 Hz (depending on motor and load) should be avoided in V/f control.

A motor can still be operated well into the field weakening region, up to about 70 Hz, if torque reduction is taken into account.

When operating a motor above the rated frequency, one should generally make sure that the increased speed does not overload mechanical components such as ball bearings, seals or gears.

Fan-cooled geared motors in particular become more inefficient with increasing speed as the more rapidly rotating fan wheel generates greater air resistance.

V/f regulated variable frequency drives can be used in applications where the speed of an asynchronous motor needs to be adjusted and where there are no precise requirements for speed stability and accuracy.

The accuracy of the rotational speed of an asynchronous motor connected to a V/f controlled variable frequency drive is approx. ± 1–5%.

Theoretically, several asynchronous motors can be operated in parallel on V/f controlled variable frequency drives if the variable frequency drive is large enough to supply the motor current for several motors.

4.1.2   Sensorless control variable frequency drives

Sensorless control variable frequency drives are becoming more efficient and cheaper as microprocessors keep getting better and faster.

A sensorless control variable frequency drive also operates according to the principle of the characteristic V/f curve, but there is additional communication between the variable frequency drive and the motor.

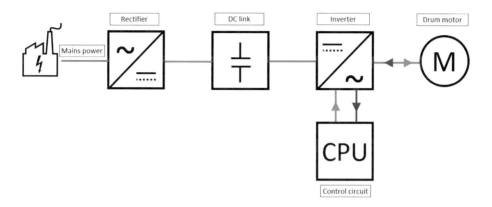

Fig. 4.7 Block diagram, variable frequency drive with sensorless control

As the name suggests, that communication takes place without sensors.

The sensorless control works with both asynchronous and synchronous motors. However, the type of motor technology (asynchronous or synchronous) must be set when setting the parameters of the sensorless control variable frequency drive.

Not all sensorless control variable frequency drives are able to control permanent magnet synchronous motors. One should therefore always check in advance whether the sensorless variable frequency drive can operate the motor technology used.

The sensorless communication between the frequency converter and the motor takes place primarily via a precise current measurement. Based on the motor current that is measured, the sensorless variable frequency drive can calculate the rotor position and motor speed fairly accurately by means of complex calculations in the control circuit.

In sensorless control, the communication between the variable frequency drive and the motor takes place via the motor power supply cable. (See Fig. 4.7.)

The actual motor speed is compared with the parameterised motor setpoint speed in the control circuit. If the actual speed deviates from the setpoint speed, then the variable frequency drive takes appropriate countermeasures and steadily ensures a nearly constant motor speed, which hardly varies even during load changes.

The sensorless control has a similar function to the cruise control in a car.

The cruise control ensures that the car runs at a constant speed, regardless of whether the car is driving along a flat stretch or on an incline.

Since sensorless control requires precise current measurement, care must be taken when dimensioning the variable frequency drive to ensure that the sensorless control variable frequency drive also matches the motor current.

If the selected variable frequency drive is too large, the motor current can no longer be determined accurately enough and the sensorless control no longer functions properly.

For the sensorless control of permanent magnet synchronous motors, in addition to precise motor current measurement, the induced voltage fed back by the synchronous motor is also analysed. This allows the rotor position to be determined even more accurately.

Since sensorless control variable frequency drives became affordable, permanent magnet synchronous motors have been playing an increasingly important role in industrial applications.

When connected to a permanent magnet synchronous motor, some sensorless variable frequency drives can generate incremental encoder signals from the sensorless control. These signals can be used for simple positioning applications.

The control accuracy for sensorless control is approximately 0.05%.

The accuracy depends very much on the hardware and the parameter settings of the variable frequency drive.

The setting of parameters for variable frequency drives with sensorless control must be very accurate. Incorrect parameters can have a negative impact on the control.

The effects are then most apparent in the motor, where overheating or loud mechanical noises may be noticed.

The most important parameters that must be set for sensorless control are:

General:
The *motor's rated voltage and rated frequency* to determine the vertex.
The *motor's rated current* to set the motor protection correctly.
Number of motor poles and nominal rotor speed to determine the rotational speed.

For permanent magnet synchronous motors:
Motor back-EMF VKE – *is often stated in V/krpm*
Current controller – *the current controller can influence the load behaviour of the motor*
Speed controller – *the speed controller can influence rotary speed behaviour.*

The parameters for the current and speed control must be adjusted if the motor reacts to a speed or load change either too strongly or too weakly.

If the motor responds too strongly, then this is called hard control. If the motor reacts too weakly, then this is called soft control.

How to best adjust the controllers always depends on the application.

Sensorless control ensures that the motor behaves very powerfully and with stable speed over the entire control range.

With sensorless control, asynchronous motors can even be operated at frequencies well below 20 Hz.

Synchronous motors have become increasingly interesting thanks to sensorless technology, as there is now a low-cost way to operate synchronous motors.

Since the sensorless variable frequency drive must communicate with the motor, only one motor can be used per sensorless control variable frequency drive.

It is not possible to operate more than one motor on a sensorless control variable frequency drive.

4.2   Encoder systems

If the sensorless control is still too inaccurate, then one has to mount a so-called rotary encoder on the rotor shaft of the asynchronous or synchronous motor.

Rotary encoders deliver data on the respective rotor position and rotational speed for every rotor revolution.

Rotary encoders may be subdivided into incremental and absolute encoders and then there is also the conventional resolver.

Absolute encoders are displacement sensors that can indicate the exact position over a longer distance (multiturn) without having to conduct a reference run after switching on a system.

Absolute encoders are not currently used in drum motors and are therefore not further discussed in this book.

Encoders provide precise information about the instantaneous behaviour of the rotor. They are required for speed control in servo systems or in positioning applications.

Positioning applications are applications which are used, for example, where a conveyor belt must always stop precisely at a specific position.

In order to always find this position again, it is necessary to define the travel path of the conveyor belt or the drum motor.

A rotary encoder provides the information needed to be able to do this positioning accurately.

Example:

In a cake factory, cherries should automatically be placed in the centre of the cupcakes.

The cakes are transported via a conveyor belt with drum motor drive and integrated rotary encoder.

The signals of the rotary encoder are counted in an electronic counter.

The distance from cake to cake is 18 encoder pulses in this example.

After every 18 encoder pulses detected by the electronic counter, the command is given to briefly stop the drum motor and drop a cherry.

In this way, the cherries always fall in the centre of the cupcakes. (See Fig. 4.8.)

4.2.1   Incremental encoders

In the technical sense, "incremental" means "stepwise" measurement or movement.

An incremental encoder generates a digital signal for a specific displacement, or for a rotational movement, for a specific angle.

Most incremental encoders work with 5 or 24 V.

0 V = digital 0

5 or 24 V = digital 1

The displacement for one signal level change is a fixed definition in the encoder.

The time elapsed from one signal level change to the next is defined by the speed of the rotor. (See Fig. 4.9.)

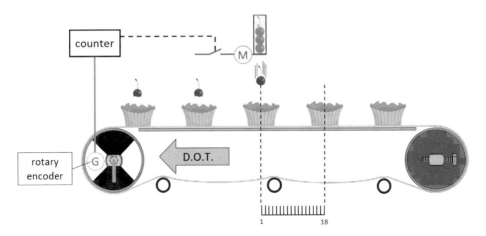

Fig. 4.8 Simplified example of a positioning application

Fig. 4.9 Incremental signal

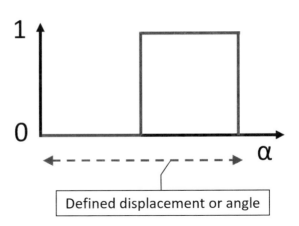

If the displacement for one increment is known, then it is only necessary to count the increments in order to be able to determine the distance travelled or the angle.

An incremental encoder can be set up using a simple transistor circuit, for example. A transistor is an electronic switching device that can switch a voltage or current on/off very rapidly and without contact.

A signal disc integrated, for example, into a ball bearing, switches the transistor on and off at defined intervals. (See Fig. 4.10.)

The resolution, meaning the accuracy of the encoder, is dependent on the magnitude of displacement between turning the transistor on or off.

The smaller the resolution, the more increments can be achieved in the encoder per revolution.

Common encoder resolutions, also called line counts, are 32, 64, 128, 256, 512, 1024, 2048, and 4096 increments per rotor revolution.

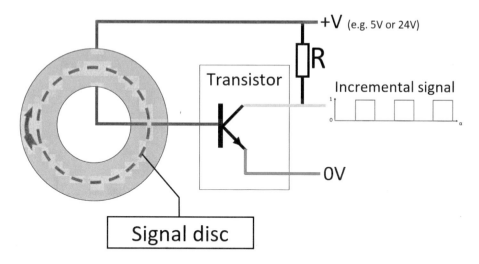

Fig. 4.10 Simplified representation of the function of an incremental encoder

However, when selecting the resolution, attention must be paid to the specifications and counting speed of the electronic diagnostic system (e.g. encoder module in the VFD or Programmable Logic Controller, PLC).

Example:
Rotor speed [nr] = 3000 rpm
Encoder resolution [INC] = 1024 /rev (increments per revolution)
Signal frequency [fENC] = ?

Formula:

$$f_{ENC} = \frac{nr}{60\,s} * INC$$

$$f_{ENC} = \frac{3000\,rpm}{60\,s} * 1024/rev$$

$$f_{ENC} = 51.2\,kHz$$

In the example above, the signal input of the diagnostic electronics must be able to count at a minimum speed of 51.2 kHz in order to evaluate the encoder signal correctly.

The electronic diagnostic system can accurately determine the rotational speed of the rotor from the signal frequency and line count.

In the example calculated above, the encoder generates 1024 increments per revolution. If the gear ratio of the drum motor is known, then the number of increments generated when the drum of the drum motor has completed one full revolution can be calculated.

Example:

Encoder resolution [INC] = 1024 /rev (increments per rotor revolution)

Gear ratio [i] = 25

Resolution, translated to the drum [INCDR] = ?

Formula:

$$INC_{DR} = INC * i$$
$$INC_{DR} = 1024/rev * 25$$
$$INC'_{DR} = 25,600/rev$$

In the example, the rotor must turn 25 times before the drum has completed one full revolution.

This results in 25,600 increments per drum revolution.

The circumference of the drum motor thus corresponds to the feed rate of the conveyor belt.

Example:

Drum diameter [d] = 81.5 mm

Resolution, translated to the drum [INCDR] = 25,600/rev

Drum circumference [C] = ?

Formula:

$$C = d * \pi$$
$$C = 81.5\,mm * \pi$$
$$C = 256\,mm$$

In the example, the circumference of the drum and thus the linear belt feed per drum rotation is about 256 mm.

One drum revolution corresponds to 25,600 increments in this example. This means 256 mm linear belt feed = 25,600 increments.

This again corresponds to 100 increments per 1 mm.

For example, if the conveyor is to be advanced by 1000 mm, then the electronic diagnostic system must count down 100,000 increments (100 inc/mm × 1000 mm) and then send the motor a stop command.

A minor disadvantage, which is not of great importance in applications with drum motors, however, is the delay time between the individual signals.

Between signal level changes, the encoder remains at level 1 or at level 0 for a brief moment. The diagnostic electronics cannot determine the exact instantaneous position of the rotor for this brief moment.

An encoder which is to be used for direct control of the motor must therefore have a sufficiently high resolution.

As the position of the rotor cannot be determined absolutely by means of incremental encoders, a system with an incremental encoder usually has to be referenced each time it is switched on.

Fig. 4.11 Signal of an incremental encoder with A and B track

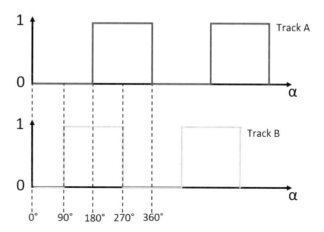

The higher the line count, the more accurate the positioning can be. If the resolution of the incremental encoder is high enough (e.g. 1024 /rev), then it may even be sufficient to control a synchronous motor.

Low resolution encoders (e.g. 32 /rev) are also often evaluated directly via a PLC or a suitable digital input from the VFD.

The electronic counter then only controls the start and stop commands for the positioning.

An encoder can thus be used to determine the speed and the position of a motor.

Another data point that is needed for an automated process, is information about the direction of rotation.

If an encoder only has one signal track, the diagnostic electronics cannot determine the direction of rotation.

Most encoders therefore have at least one additional signal track, which is usually offset by 90° to the first signal track. (See Fig. 4.11.)

The second, offset signal track is used to determine the direction of rotation by evaluating which of the two signal tracks first jumps to level 1.

Incremental encoders have the advantage that a digital signal is easy to evaluate.

There are many ways to process digital incremental signals,

which is why incremental encoders are very widespread in industry.

An incremental encoder is usually an electronic component, so it can sometimes get damaged by incorrect handling.

Typical pitfalls include:
- Reversing the polarity of the input voltage (+ and – have been switched).
- Short circuit at the signal output. Either between different signals or to ground.
- Accidental operation at too high an input voltage.
- Overheating due to high ambient temperature.
- Non EMC-compliant wiring.

Some encoders have reverse polarity protection or overload protection, while other encoders do not.

Occasionally, barely measurable, brief overvoltage spikes can damage a digital encoder. If an encoder fails despite proper wiring, it is usually very difficult to determine the exact cause of the error.

4.2.2   Resolvers

A resolver is a rotary encoder that outputs an analog signal. Unlike with an incremental encoder, the exact rotor position and angular position can be determined absolutely from the analog signal of a 2-pole resolver within one rotor revolution.

In order to understand how the resolver functions in detail, one needs to understand the principle of electromagnetism and electromagnetic induction.
A resolver theoretically works just like a transformer, except that the primary winding rotates.

Fundamental principles:
Transformer principle:
A transformer is an electrical component with an input coil (primary winding) and an output coil (secondary winding).
If an alternating voltage is connected to the primary winding, a magnetic field is created according to the principle of electromagnetism.
The magnetic field of the primary winding induces a voltage in the secondary winding, according to the principle of electromagnetic induction.
The primary and secondary windings are not usually electrically connected to each other. The magnitude of the induced voltage is affected by the number of turns in the primary and secondary windings, respectively. (See Fig. 4.12.)

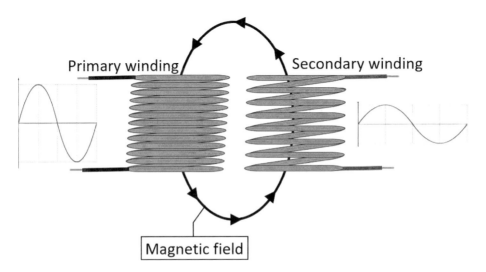

Fig. 4.12 Simplified representation of a transformer

A resolver is composed of a stator and a rotor.

The stator consists of an excitation winding and two stator windings spatially offset by 90°.

If one compares the resolver to a transformer, the two stator windings can be considered as secondary windings.

The stator windings are also referred to as sine and cosine output.

A high-frequency AC voltage is connected to the excitation winding.

The higher the frequency of the input AC voltage, the more efficiently a transformer or resolver will operate.

An exemplary common excitation voltage for conventional resolvers would be 7 Vac with a frequency of 5 kHz–10 kHz.

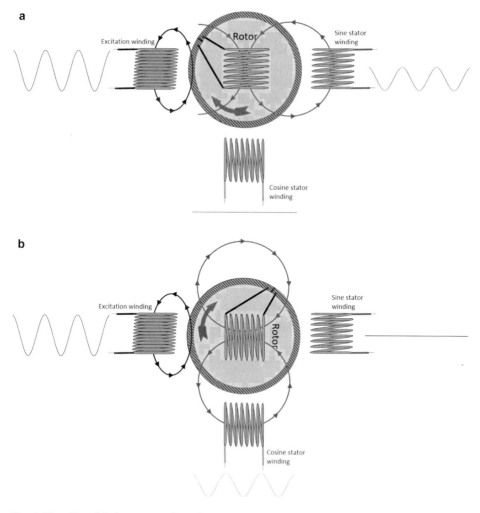

Fig. 4.13 a Simplified representation of a resolver in sine position. **b** Simplified representation of a resolver in cosine position

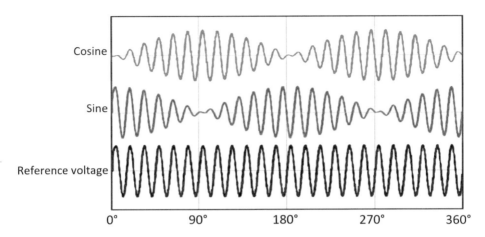

Fig. 4.14 Typical signal waveform of a 2-pole resolver from 0°–360

The alternating voltage causes an electromagnetic field to be created in the wound copper wire of the excitation winding due to the principle of electromagnetism.

Similar to a transformer, according to the principle of electromagnetic induction, an alternating voltage is induced, without contact, in a copper coil in the rotor of the resolver. The induced voltage in the rotor is in turn passed through a further excitation coil, whereby an electromagnetic field is generated once again.

Because the rotor can rotate, the rotor's rotor winding rotates at different angles to the two stator windings.

As a result, different voltage levels are induced in the two stator windings, depending on the angular position. (See Fig. 4.13a and b)

The exact position of the rotor can be determined at any time based on the magnitude of the voltages and the phase sequence, the alternating voltages in the stator windings. (See Fig. 4.14.)

Resolvers are very robust as, in principle, they consist only of coiled copper wires.

Resolvers are ideal for controlling synchronous drum motors because they can determine the exact rotor position absolutely.

In addition to controlling the motor, the data provided by the resolver can also be used for positioning.

4.3   Servo controller with encoder feedback

Encoders or resolvers are no use at all if their data and signals cannot be properly processed and implemented.

A servo controller can process the data from a rotary encoder and at the same time control the motor accordingly.

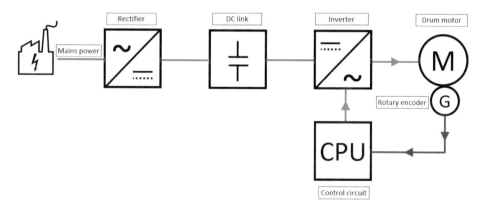

Fig. 4.15 Block diagram, servo VFD with closed control loop

In principle, servo controllers operate like sensorless control variable frequency drives, but with the big difference that the servo controller obtains real, measured data about the rotor position and speed from the encoder. Thanks to the real-time measurement by means of rotary encoders, a servo controller with encoder feedback is many times more accurate and more efficient than, for example, a sensorless control VFD.

The servo system consisting of the control circuit, inverter, motor and encoder forms a closed loop. (See Fig. 4.15.)

The higher accuracy allows the servo system to control the synchronous drum motor even more dynamically, allowing for shorter start and stop ramps. The speed control range becomes greater.

Synchronous drum motors can still be operated powerfully at, for example, 2 Hz with a servo controller with encoder feedback. This would not be possible with sensorless control.

The encoder feedback allows the motor to be used for applications that require positioning.

The speed can be controlled extremely accurately and thus does not change even when the load changes.

The combination of a synchronous motor with rotary encoder and servo inverter enables the user to get the most from the motor. There is not much else that can be achieved within the physical limits of the motor.

The highly dynamic processes of automated applications frequently use servo inverters as they not only have to communicate with the motor but also with machines or servo inverters located upstream and downstream.

In this case, the encoder information is also used to synchronise upstream and downstream operations.

However, a servo system is also associated with significantly higher hardware costs and more complicated wiring and programming.

This makes the system quite expensive in comparison.

While the control accuracy for sensorless control is about 0.05%, closed-loop control achieves an accuracy of about 0.01% or better.

Servo controllers with encoder feedback make little sense for simple applications that do not require precise positioning, synchronisation of processes, extremely low frequency control ranges or perfect speed control, because of the acquisition costs and the additional effort involved.

Drive dimensioning with drum motors

5

Drum motors are highly versatile, but traditionally they are used as belt drives in conveyors. The design of a conveyor belt can be quite complex. There are many different types of conveyor belts, conveyor designs, environmental conditions, goods to be conveyed, applications, etc., and each parameter can have a major impact on the belt drive.

There are certain rules that must be observed when dimensioning a drum motor for a conveyor belt application.

5.1   Environmental conditions

When designing and dimensioning a conveyor belt, the first thing to do is to consider the environmental conditions.

The following environmental conditions need to be assessed:
- Is it a wet, damp or dry environment?
- What is the ambient temperature?
  - maximum ambient temperature?
  - minimum ambient temperature?
- What is the altitude of the application above sea level (AMSL)?

Wet environment:

The drum motor's high degree of protection means it is predestined to be used in harsh and aggressive environments.

The ambient conditions impact on the choice of materials that are to be used with the drum motor.

In general, the harsher and more aggressive the environment, the greater the demands placed on the materials.

© Springer-Verlag GmbH Germany, part of Springer Nature 2020
S. Hamacher, *The Drum Motor*,
https://doi.org/10.1007/978-3-662-59298-4_5

Fig. 5.1 Drum motors undergoing regular cleaning (Source: Interroll.com)

Wet environments are often found in open food processing.

Where there is open food, the conveyor must be cleaned regularly, sometimes several times a day. (See Fig. 5.1)

As a rule, a lot of water and chemical cleaning agents are used. The drum motor must therefore not only be tightly sealed to prevent the ingress of water, but also be able to withstand the aggressive and sometimes slightly corrosive cleaning agents.

Particularly in applications in which the drum motor can come into contact with seawater, the materials used must be of the highest quality.

If drum motors are designed, for example, with aluminium covers, for reasons of cost, then the aluminium parts can become corroded if they come into contact with seawater and cleaning agents.

For practically all food applications involving regular cleaning, the metal parts of a drum motor should therefore preferably be made of stainless steel with the specifications AISI 303-304 or AISI 316.

But it is also important that other parts like seals, cables, insulating materials, rubber coatings, etc. are appropriate for use with cleaning agents.

Damp environment:

In damp applications marked by high humidity, or where a few splashes of water can occasionally come in contact with the drum motor, but where the drum motor is not usually wet, it is sufficient to install a drum motor with aluminium covers to save costs, if other environmental conditions will allow.

Fig. 5.2 Drum motors for baggage drop-off in airports (Source: Interroll.com)

The drum shell and the protruding shaft ends of the drum motor should be made of stainless steel to avoid rust.

Examples of damp applications include the further transport of packaged food or applications where conveyor belts are operated outdoors.

Dry environment:

Typical dry applications include applications in postal and airport logistics. (See Fig. 5.2) Dry logistics applications are frequently very cost-driven, often using drum motors with steel drums, steel shafts and aluminium lids.

Minor rust on the shafts is not a problem as there are no hygiene requirements.

Just as a rolling stone gathers no moss, the bare steel drum does not develop any rust in the area of the conveyor belt during operation.

Even if a motor should start rusting during longer downtimes, this would be rubbed off again after the belt has been in operation for a short time.

It is possible to galvanise a bare steel tube. The zinc plating is however usually worn away over time by the friction of the conveyor belt, so that the bare drum shell reappears beneath the belt.

The purpose of galvanising the drum tube is far more for rust protection during transportation, when drum motors have to be transported by sea to distant countries.

Ambient temperatures:
It is essential to know the ambient temperatures when dimensioning a drive system.

The ambient temperature for drum motors in standard operation is between +5 °C and +25 °C.

If positively driven conveyor belts are used within this temperature range, an operating reserve of at least 10% should be taken into account.

At ambient temperatures between +25.1 and 40 °C, drum motors should only be used in conjunction with friction-driven conveyor belts to allow sufficient heat to be drawn from the shell even at the higher ambient temperature.

Furthermore, drum coatings thicker than 8 mm should be avoided, as coatings that are too thick can act as thermal insulators.

The performance of the drum motor can be affected if the ambient temperature is too high. In general, asynchronous drum motors should not be used for ambient temperatures above +40 °C.

It may be possible under certain circumstances to operate a synchronous drum motor in areas warmer than +40 °C, but it is imperative to consult the drum motor supplier.

At low ambient temperatures below +5 °C down to −25 °C, one should always avoid mounting terminal boxes of any kind directly on the drum motor, as the combination of heating by the drum motor and the cool environment may cause condensation to occur in the terminal box.

Since electricity and water are known for their incompatibility, in the worst case a short circuit can even occur in the terminal box.

However, even a design with a permanently mounted motor cable is hazardous at low temperatures. In principle, any mechanical loads on the cable or movement of the cable should be avoided.

Cables with PVC insulation are particularly unsuitable for very low temperature applications, as the PVC insulation can break easily.

For this reason, cables with PUR insulation are frequently used in the low-temperature sector.

When choosing the oil for drum motor applications in the range of +5 °C to −25 °C, one should also pay attention to its properties at low temperatures. Mineral oils are often less suitable than synthetic oils.

The wrong oil may begin to solidify during longer downtimes at low temperature.

This increases the friction in the mechanically lubricated components, such as in the gear mechanism and in the ball bearings.

If the mechanical friction is too great, the drum motor can in the worst case no longer start up without assistance.

In order to prevent the seals freezing and the motor oil from solidifying, drum motors must be heated slightly during downtimes in environments below +5 °C.

In the case of a synchronous motor, this task can be performed by the VFD by it simply continuing to supply the synchronous drum motor with power, so that the motor maintains its position at standstill. The flow of current in the synchronous motor winding is enough 'to warm the motor oil sufficiently and keep it fluid. This can also prevent the seals from freezing up.

A simple trick can be used for the internal heating of asynchronous motors.

If the asynchronous drum motor is disconnected from the mains or the VFD, one can simply connect a DC voltage to any two phases of the asynchronous motor.

The correct DC voltage can vary from stator to stator because different stators may have different resistances.

Drum motor manufacturers generally indicate the DC voltage for the standstill heating system on the motor nameplate.

The DC voltage causes current to flow through the copper windings of the asynchronous motor.

However, the resulting magnetic field is static, so no voltage can be induced in the squirrel cage rotor. The asynchronous motor is therefore not made to rotate.

The DC current in the winding does however generate heat. This heat keeps the motor oil at temperature and no ice crystals can form on the seals.

Applications above 1000 m:

Technical equipment operated at altitudes higher than 1000 m above sea level must be dimensioned with a higher operating reserve on account of the lower air pressure.

As a rule of thumb, for altitudes over 1000 m one should reckon with about 1% additional operating reserve per 100 m.

5.2   It all comes down to friction (general conveyor dimensioning)

The basic calculation of a conveyor belt drive begins with the dimensioning of a lifting drive.

When dimensioning a lifting drive, one only has to calculate the force of gravity pulling on the load to be lifted.

Example:

A crane drive is to lift a pallet carrying 150 kg cement.

The total mass [m] is 150 kg cement + 35 kg pallet = 185 kg (See Fig. 5.3)

The acceleration due to gravity [g] is approx. 9.81 N/kg. This means that gravity pulls on the load to be lifted with a force of 9.81 N per kg of its mass.

This results in the following formula for the force [F]:

Formula:

$$F = m * g$$
$$F = 185\,kg * 9.81\,N/kg$$
$$F = 1814.85\,N$$

In order to lift the pallet, the crane drive must deliver a force [F] of at least 1814.85 N.

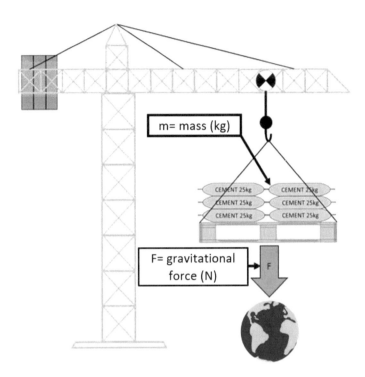

Fig. 5.3 Dimensioning a lifting drive

But the force [F] alone does not say anything about the power [P] of the drive.
To calculate the mechanical power [Pmech] you need the speed [v], with which the load
[m] is to be pulled upwards.
In our example, the customer requires that the load [m] be pulled upwards at a speed [v]
of 0.5 m/s.
The mechanical power [Pmech] of the drive is the product of force [F] and speed [v].

Formula:

$$P_{mech} = F * v$$
$$P_{mech} = 1814.85\,N * 0.5\,m/s$$
$$P_{mech} = 907.43\,W$$

The lifting of a load is the most demanding application for a drive, as it acts in opposition
to the total gravitational force of the load to be lifted.
The gravitational pull thus opposes the force of the crane drive by 180°.
However, if the same load is to be moved horizontally, less force is usually required for
this than for lifting.
As everyone knows, it is practically impossible to pick up a car with ones bare hands.

But if one pushes a stationary car with the handbrake released on a horizontal road, one can easily set the heavy car in motion.

For the horizontal movement, the gravitational pull is no longer 180° in opposition, but only at 90°.

The force needed to move the car horizontally depends very much on the friction between the road and the car.

When the handbrake is released, the wheels of the car, which are mounted on ball-bearings, can rotate easily. The so-called rolling friction of the ball bearings is very low. That's why relatively little force is needed to get the car moving.

When the handbrake is applied, the car's wheels are locked. In order to set the car in motion when its wheels are locked, much more force would be necessary, since the much greater friction between the rubber of the tyres and the rough road surface must now be overcome. The efficiency with which one can move something horizontally is therefore dependent on the friction factor between the load to be moved and the surface over which the load is to be pushed or pulled.

The art of dimensioning conveyors lies in determining or calculating the friction that arises and the forces needed to overcome this friction.

The greatest friction occurs between the conveyor belt and the surface on which the conveyor belt runs or is carried.

The friction is dependent on the material of the underside of the conveyor belt, the material of the upper surface of the conveyor bed and whether the carrying surface is a slider bed or is equipped with rollers.

The lower the friction between conveyor belt and conveying surface, the more efficient a conveyor will be.

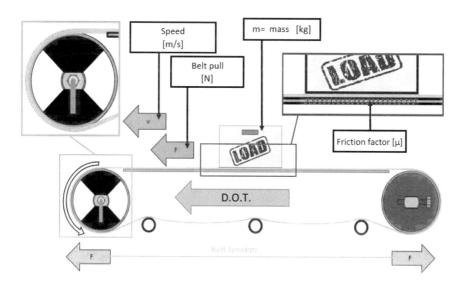

Fig. 5.4 Friction, forces and speed

Tab. 5.1 Common friction factors in conveyor technology

Belt material	Slider bed material		Rolling friction
	PE	Steel/stainless steel	
PE	0.3	0.15	
PP	0.15	0.26	
POM	0.1	0.2	0.05
PVC/PU		0.3	
Polyamide or polyester		0.18	
Rubber	0.4	0.4	

But friction is not always undesirable. The drive drum requires friction. The friction between conveyor belt and drive drum must always be greater than the friction between conveyor belt and conveying surface.

If the friction on the drive drum is too low, it will spin. (See fig 5.4)

In order to calculate the force needed to overcome the friction between conveyor belt and conveying surface, the so-called friction factor is required. (See Tab. 5.1) As a rule, the friction factor is less than 1.

A lifting application would have a friction factor of 1.

Once the coefficient of friction has been determined, the force required for the horizontal movement may be calculated.

Simply multiply the friction factor $[\mu]$ by the force that would be required to lift the load.

Example:

A pallet carrying 150 kg of cement is to be conveyed horizontally on a conveyor belt with a PVC belt and a sliding conveying surface that has a steel slider bed.

The PVC conveyor belt weighs 2 kg/m^2. The conveyor belt is 0.6 m wide and the distance between the shafts of the idler pulley and the drum motor is 2.5 m. (See Fig. 5.5)

It is not only the load that rests on the sliding bed, but also a portion of the conveyor belt.

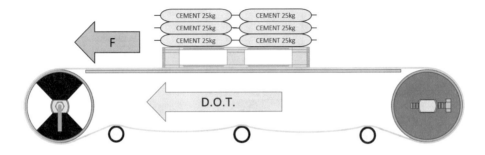

Fig. 5.5 Example of drive dimensioning for a horizontal conveyor

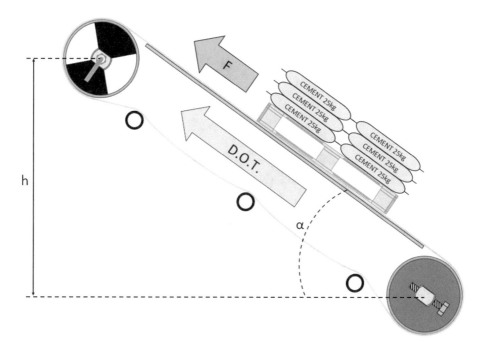

Fig. 5.6 Example of drive dimensioning for an ascending conveyor

The mass of the conveyor belt in the upper strand must therefore also be taken into consideration.

Since the conveyor belt is usually run on low-friction rollers in the lower strand, the mass of the belt can be neglected here.

The mass of the conveyor belt in the upper strand may be calculated as follows:
Conveyor length axis-to-axis (AA) [L] = 2.5 m
Conveyor belt width [BW] = 0.6 m
Specific mass of conveyor belt [mb] = 2 kg/m²
Belt mass in upper strand [mbu] = ?

Formula:

$$m_{bu} = L * BW * m_{b}$$
$$m_{bu} = 2.5\,m * 0.6\,m * 2\,kg/m$$
$$m_{bu} = 3\,kg$$

The total mass [m] is 150 kg cement + 35 kg pallet = 185 kg.
The conveyor belt in the upper strand weighs 3 kg [mbu].

Mass to be conveyed [m] = 185 kg
Conveyor belt in upper strand [mbu] = 3 kg
Gravitational acceleration [g] = 9.81 N/kg
Friction factor [μ] = 0.3
Belt pull [F] = ?

Formula:

$$F = (m + m_{bu}) * g * \mu$$
$$F = (185\,kg + 3\,kg) * 9.81\,N/kg * 0.3$$
$$F = 553.28\,N$$

The smaller the friction factor, the less the force required to move the load.
So it all comes down to friction.
Now the power required for the conveyor belt drive can be calculated.
For this one needs to know the belt speed, which is usually given for the application.

Example:
Conveying speed [v] = 0.5 m/s
Required belt pull [F] = 553.28 N
Mechanical power [P_{mech}] = ?

Formula:

$$P_{mech} = F * v$$
$$P_{mech} = 553.28\,N * 0.5\,m/s$$
$$P_{mech} = 276.64\,W$$

Friction factors can change depending on the application, the goods to be transported or the environment.
In wet and damp applications, for example, the conveyor belt may be strongly attracted to the slider bed by adhesion forces. The friction factor is then very large, especially during the conveyor's start-up phase.
    Another example often occurs in open meat production.
Animal fats, which may spread over the conveyor during meat processing, improve the sliding properties and can thus potentially improve the friction factor.
    Conversely, in cheese production, for example, cheese particles between the conveyor belt and the slider bed can have a negative effect on the coefficient of friction.
    One should also expect a higher friction factor in freezers, as the conveyor belt is less flexible here due to the cold, so more power is likely to be needed.

A bit of experience and intuition is needed to correctly assess the conveyor's friction factor, especially in food applications.

Basically, it is always a good idea to add an operating reserve of at least 20% or more to the belt pull calculated for the drive.

Formula:

Additional operating reserve of 20% [F_{res}]

$$F_{res} = F * 1.2$$

As already described in section 5.1 on environmental conditions, an additional operating reserve must be taken into account when installed at an altitude of more than 1000 m above sea level owing to the reduced air pressure.

Example:

A belt pull of 500 N [F] and motor power [P_{mech}] of 300 W were calculated for an application located at less than 1000 m altitude.

This motor is now to be operated at an altitude of 2000 m above sea level.

   Altitude above sea level in m [alt] = 2000 m

Calculated belt pull [F] = 500 N

Calculated mechanical motor power [P_{mech}] = 300 W

Additional operating reserve required per m above 1000 m = 0.01 $\frac{\%}{m}$

Reserve factor above 1000 m [f_{alt}] = ?

Formula (only applies from 1000 m above sea level):

$$f_{alt} = 1 + \frac{(alt - 1000\,m) * 0.01\frac{\%}{m}}{100}$$

$$f_{alt} = 1 + \frac{(2000\,m - 1000\,m) * 0.01\frac{\%}{m}}{100}$$

$$f_{alt} = 1.1$$

Since the motor becomes weaker if it is higher than 1000 m above sea level, the belt pull [F] and mechanical motor power [P_{mech}] that have been dimensioned must be multiplied by the reserve factor [f_{alt}].

Formula (only applies from 1000 m above sea level):

Belt pull at >1000m above sea level [F_{alt}]

$$F_{alt} = F * f_{alt}$$
$$F_{alt} = 500\,N * 1.1$$
$$F_{alt} = 550\,N$$

If a belt pull of 500 N was sufficient for an application located below 1000 m, a belt pull of 550 N would be required for the same application when installed at a location 2000 m above sea level.

Formula (only applies from 1000 m above sea level):
Mechanical power at >1000m above sea level [P_{alt}]

$$P_{alt} = P_{mech} * f_{alt}$$
$$P_{alt} = 300\,W * 1.1$$
$$P_{alt} = 330\,W$$

Since more belt pull [F_{alt}] is required at the same belt speed [v], logically, the mechanical power [P_{mech}] must increase as well.

5.2.1   Special characteristics of ascending and descending conveyors

Ascending or descending conveyors are in principle a combination of a crane drive and horizontal conveyor drive. Here, in addition to the friction factor, one must also consider the opposing gravitational force, which depends on the angle of inclination.

Ascending conveyor example:
A pallet carrying 150 kg of cement is to be conveyed upwards on a conveyor belt with a PVC belt and sliding conveyor surface made of steel at a **gradient** of α = 35°.
The total mass [m] is 150 kg cement + 35 kg pallet = 185 kg.
The conveyor belt in the upper strand weighs 3 kg [mbu].
Mass to be conveyed [m] = 185 kg
Conveyor belt in upper strand [mbu] = 3 kg
Gravitational acceleration [g] = 9.81 N/kg
Friction factor [μ] = 0.3
Slope [α] = 35°
Belt pull [F] = ?

Formula:

$$F = (m + m_{bu}) * g * \mu + (m + m_{bu}) * g * \sin(\alpha)$$
$$F = (185\,kg + 3\,kg) * 9.81\,N/kg * 0.3 + (185\,kg + 3\,kg) * 9.81\,N/kg * \sin(35°)$$
$$F = 1611.12\,N$$

Example:
Conveying speed [v] = 0.5 m/s
Required belt pull [F] = 1611.12 N
Mechanical power [P_{mech}] = ?

Formula:

$$P_{mech} = F * v$$
$$P_{mech} = 1611.12\,N * 0.5\,m/s$$
$$P_{mech} = 805.56\,W$$

Descending conveyors:
In theory, one would only need to negate the sign of the gradient angle to calculate the driving force required for a descending conveyor.
The theoretically required driving force is less for a descending conveyor because gravitational force acts in the conveying direction.
If the gradient angle is too steep, a negative drive power may even be calculated. This is the case when the force of gravity is greater than the force necessary to overcome the friction between the conveyor belt and slider bed.

It is important to realise that the electric motor operates more as a brake than a drive in descending conveyors.
If the calculated drive power of the motor is too weak, particularly for steep angles of inclination, then the gravitational force pulls so strongly on the load that the conveying speed gets out of control. A motor that is not dimensioned powerfully enough cannot hold the load back any more.

The calculations to dimension a descending conveyor must therefore also be carried out using a *positive* angle.
If a motor can pull the load up, then the motor can also control the load's descent or slow it down.

For heavy loads and at steep angles, the conveyor belt is driven by gravity and mechanically drives the electric motor. A mechanically driven asynchronous or synchronous motor acts like a generator and produces energy.
If this energy is not dissipated, then the speed of the conveyor can get out of control.
It therefore often makes sense to include a suitable VFD with a brake chopper circuit when dimensioning a descending conveyor. The VFD regulates the speed and keeps it constant.
If the electric motor is operated as a generator, then the brake chopper circuit in the VFD can conduct the excess energy into a braking resistor.
The resulting motor or generator current slows down the conveyor belt. A controlled, electronically stable conveying speed can thus be ensured on descending slopes.

Ascending and descending conveyors with pivot points:
Ascending and descending conveyors with bends or pivot points, so-called L-frame conveyors, gooseneck conveyors or Z-frame conveyors (See Fig. 5.7), can be calculated as if they were straight ascending or gradient conveyors.
This requires the travel height which is to be overcome and the total conveyor length. The dimensions for a straight conveyor with gradient can be calculated from these two values, using the formula from the example above.

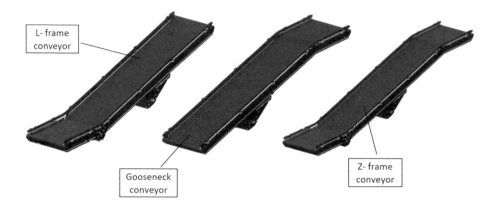

Fig. 5.7 Special shapes: ascending and descending conveyors (Source: Interroll.com)

Z-frame conveyor example:

A Z-frame conveyor is to be converted so that the formula F = m * g * μ + m * g * sin (α) can be used.

In other words, the Z-frame conveyor must be converted into a straight conveyor with a shallower gradient angle [α2] and a longer inclined section [CL2]. (See Fig. 5.8)

Horizontal top length [TL] = 1 m
Conveyor length [CL] = 3.5 m
Horizontal bottom length [BL] = 0.5 m
Height to overcome [h] = 2 m
Conveyor length, converted into a straight ascending conveyor [CL2] = ?

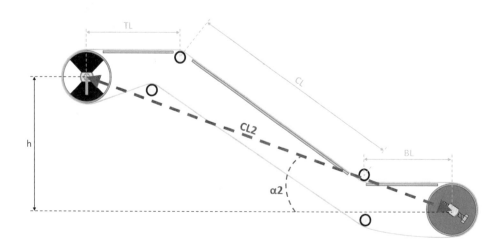

Fig. 5.8 Example drive dimensioning of a Z-frame conveyor

Formula:

$$CL2 = TL + CL + BL$$
$$CL2 = 1\,m + 3.5\,m + 0.5\,m$$
$$CL2 = 5\,m$$

The diagonal [CL2], if calculated for simplicity by adding together TL, CL, and BL, is actually a little too long. But that does not have a major impact on the dimensioning. For L-frame conveyors TL = 0 m and for gooseneck conveyors BL = 0 m.

Gradient angle, converted into a straight ascending conveyor [α2] = ?

Formula:

$$\alpha2 = \tan^{-1}\left(\frac{h}{CL2}\right)$$
$$\alpha2 = \tan^{-1}\left(\frac{2\,m}{5\,m}\right)$$
$$\alpha2 = 21.8°$$

Now the formula F = m * g * μ + m * g * sin(α2) can be used.

5.3   Additional frictional forces

Additional friction may arise due to components mounted on the conveyor or unusual conveyor shapes and this must be added to the calculated belt pull.

For example, cleaners, scrapers, brushes, modular belt curves and pivot points on L-frame, gooseneck or Z-frame conveyors create additional friction that is not to be underestimated. Depending on the conveyor design, the pivot points on L, gooseneck or Z-conveyors can each generate approximately 50–100 N of frictional force. (See Fig. 5.9)

In open food applications, cleaners or scrapers are often attached to the head or beneath the conveyor to remove coarse product debris from the belt surface. The cleaners must be fitted snugly against the conveyor belt with a certain amount of pressure. As a result, additional friction can arise, especially if residues of the conveyed material build up on the cleaner over time.

One can assume additional frictional forces of about 75–100 N per cleaner or scraper. (See Fig. 5.10)

Additional friction is also generated with curved modular belts. (See Fig. 5.11)

The friction arising in the bends can sometimes be quite high. Depending on the belt type and design, approximately 50–200 N of additional friction may arise at each bend in the modular belt.

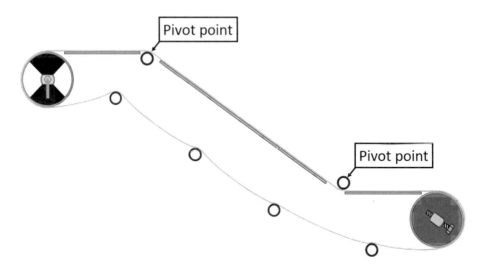

Fig. 5.9 Friction can occur at the pivot points

Fig. 5.10 Conveyor belt with drum motor and cleaner (Source: Interroll.com)

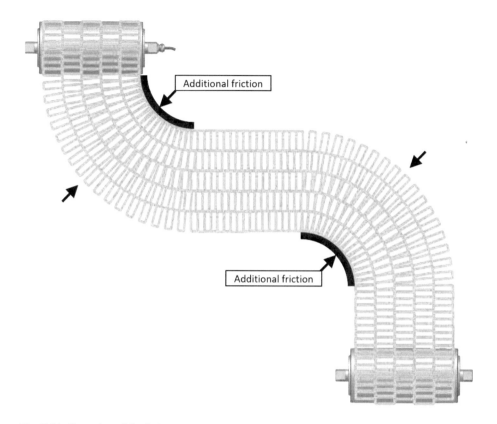

Fig. 5.11 Curved modular belt

5.4   Belt tension

A very important factor in dimensioning a conveyor with a drum motor, and one that is often underestimated, is the belt tension.

No reliable dimensioning is possible if the conveyor belt is not known.

Friction-driven conveyor belts require friction between the conveyor belt and drive drum. This friction is generated primarily through pressure by tensioning the endless belt loop around the drive drum and pulley.

The tighter the conveyor belt is tensioned, the greater the grip between the drive drum and the conveyor belt.

The forces involved can be immense. Many people have no idea how extremely high belt tension forces can be and the negative effects they can have on a diverse range of conveyor components.

All the data required for the conveyor belt can usually be found in the belt data sheet or obtained from the belt manufacturer.

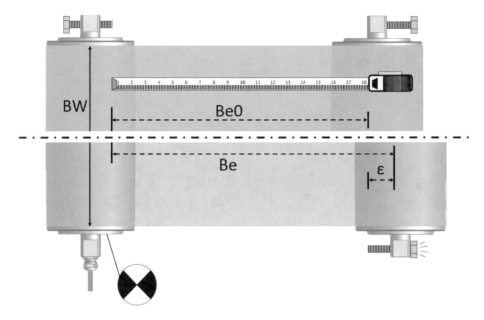

Fig. 5.12 Measurement of the belt elongation

As already explained in section 2.1, the following belt data is needed to calculate the belt tension:
- Belt width in mm [BW]
- K1% value in N/mm [K1%]

The belt elongation [ε] must also be known or determined as a percentage.
The belt elongation can be relatively easily measured while adjusting and tensioning the conveyor belt.

Determination of the belt elongation:
1. Two marks are made on the loose, untensioned conveyor belt (e.g. a line with a pen or with a strip of adhesive tape).
   The markings should be centrally positioned on the conveyor belt, since the highest belt tension and belt elongation is to be expected here due to the convexity of the drive pulley.
   The markers should be as far apart as possible. The farther apart the markers are, the more accurate the subsequent measurements will be.
   The exact distance between the two marks must now be measured. The symbol for the distance between the marks in the *unstretched state* is [Be_0].
2. Now the drum motor can be turned on. The drum motor will probably slip under the conveyor belt at first. One can then carefully increase the distance between the drum motor and the pulley until the belt is eventually picked up.

Care should be taken to tension the conveyor belt evenly on both sides to achieve a smooth belt run.

Tensioning the conveyor belt causes it to stretch a little.

Once the conveyor belt is running straight and has enough grip to move the load that is to be conveyed without the drive slipping, the drum motor can be turned off again.

3. Now the previously applied markers need to be found and the distance between the markers remeasured.

   The two markers will have moved further apart, since the belt has elongated a little as a result of the tension.

   The symbol for the distance between the marks in *the elongated state* is [Be]. (See Fig. 5.12)

4. The two measured values can now be used to calculate the percentage expansion of the conveyor belt after tensioning.

Example:

Two markings are made in the centre of an untensioned belt at a distance of exactly [Be0] = 1000 mm apart.

After the belt has been tensioned and straightened, the distance between the two markers has increased by 4 mm, to a total of [Be] = 1004 mm.

Distance between markings in unstretched condition [Be0] = 1000 mm

Distance between markings in tensioned state [Be] = 1004 mm

Extension in % [ε] = ?

Formula:

$$\varepsilon = \frac{Be * 100\ \%}{Be0} - 100\ \%$$

$$\varepsilon = \frac{1004\ mm * 100\ \%}{1000\ mm} - 100\ \%$$

$$\varepsilon = 0.4\ \%$$

Once the belt elongation [ε] in % is known, the K1% value of the conveyor belt is also needed. The K1% value is usually given in the belt data sheet or can be requested directly from the conveyor belt manufacturer.

The belt width is also required. The belt width can be measured directly on site or it is specified in the drive dimensions.

Example:

A 600 mm wide conveyor belt was tensioned with 0.4% elongation. The belt data sheet specifies a dynamic and a static K1% value.

The static K1% value is usually the larger value. To be on the safe side, one should use the static K1% value in calculations. In our example the K1% value is 8 N/mm.

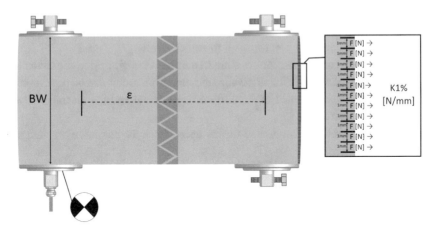

Fig. 5.13 Belt tension

Belt width [BW] = 600 mm
Percentage belt elongation [ε] = 0.4%
Belt tension force per mm belt width at 1% extension [K1%] = 8 N/mm
Belt tension force in N [TE] = ?

Formula:

$$TE = BW * K1 \% * \varepsilon * 2$$
$$TE = 600\,mm * 8\,N/mm * 0.4 * 2$$
$$TE = 3840\,N$$

The derivation of the formula is quite simple. The K1% value is typically expressed in N/mm and indicates how much force per millimetre of belt width is needed to stretch the belt by 1%. (See Fig. 5.13)

In the example, however, only a belt elongation of ε = 0.4% was measured, so K1% can be multiplied by ε.

Since the K1% value is effective per mm conveyor belt width [BW], the conveyor belt width [BW] can also be multiplied by this.

The resulting formula for belt tension is: BW * K1% * ε.

A factor of 2 is often omitted. The belt tension BW * K1% * ε refers to just one conveyor belt layer.

But since it is an endless, closed conveyor belt, there is one conveyor belt layer in the upper strand and one in the lower strand.

If the belt in the upper strand is elongated, for example, by 0.4%, then the belt also stretches by 0.4% in the lower strand.

The factor of 2 thus also takes into account the belt tension which is generated in the lower run.

In the previous example, a belt tension force of 3840 N was calculated.

3840 N corresponds to a mass of 3840 N/9.81 N/kg = 391.4 kg.

Simply tensioning the conveyor belt by a few millimetres puts a strain of several hundred kilograms on the drum motor and ball bearings.

This strain on the ball bearings has a negative impact on the ball bearing life.

5.5   Drum motors with rubber lagging or profiles for positively driven belts

Smooth rubber lagging on the drum motor is intended to increase the grip between the belt and drive pulley.

This requires less belt tension, which has a positive effect on ball bearing life.

Even more gentle on the ball bearings are positively driven belts such as modular belts or thermoplastic belts, since positively driven belts do not have to be tensioned.

However, having rubber lagging, a profile or sprocket wheels on the drum motor increase the unwind roll diameter of the belt, which increases the belt speed compared to an uncoated drum motor.

Having a larger drum diameter at constant torque means that less pulling force is available at the larger outer diameter.

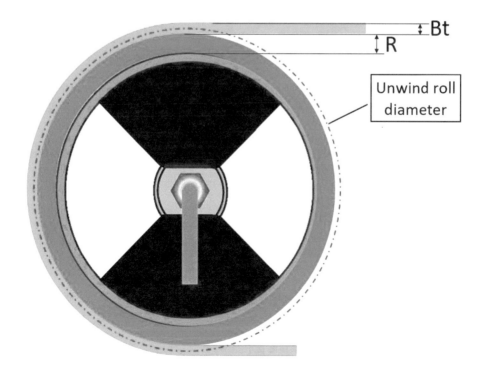

Fig. 5.14 Unwind roll diameter with friction driven belts and rubber lagging

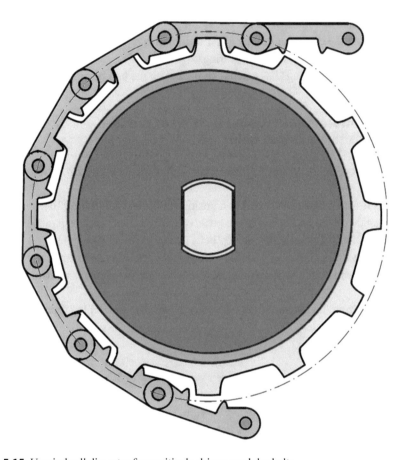

Fig. 5.15 Unwind roll diameter for positively driven modular belts

The thickness of the conveyor belt and an enlarged drum diameter must therefore be taken into account in the drive dimensioning.

Friction driven conveyor belts
Friction driven conveyor belts unroll in the middle of the belt. (See Fig. 5.14)
If rubber lagging or another coating is applied to the drum shell, the unwind roll diameter of the belt is calculated as follows:

Formula:
Thickness of the rubber lagging in mm [R]
Thickness of the belt in mm [Bt]
Diameter of uncoated drum [Ø]
Unwind roll diameter of the belt in mm [Øfinal]

$$\varnothing\text{final} = \varnothing + 2 * R + Bt$$

Positively driven modular belts

Positive drive modular belts are usually driven by means of a profile or sprocket wheels.
The unwind roll diameter is called the pitch circle diameter for modular belts.
This is often abbreviated to PCD.
The PCD is an imaginary circle that runs through the centre of the hinges when the modular belt is unrolled.
The PCD is usually given by the profile or sprocket wheel manufacturer.
For modular belts, the unwind roll diameter corresponds to the pitch circle diameter [PCD].
(See Fig. 5.15)

Formula:
Pitch circle diameter in mm [PCD]

$$\varnothing \text{final} = PCD$$

Positively driven thermoplastic belts

Positively driven thermoplastic belts are pushed along on the interlocking teeth.
The conveyor belt rolls over the top of the profile or sprocket wheels.
The outer diameter [OD] of the profile is usually given by the profile or sprocket wheel manufacturer or, if it is at hand, it can easily be measured.
In the case of positively driven thermoplastic belts, the unwind roll diameter corresponds to the outer diameter [OD]. (See Fig. 5.16)

Formula:

$$\varnothing \text{final} = OD$$

The larger unwind roll diameter [Øfinal] must now be converted to the drum motor catalogue values.
Compared to the catalogue data, the belt speed increases and the belt pulling force decreases with a larger unwind roll diameter.
The unwind roll diameter [Øfinal] can now be used to calculate the correction factors for conversion to catalogue values.

Example:
The mechanical values in a drum motor catalogue refer to an uncoated drum diameter [Ø] of 138 mm.
A modular belt profile changes the unwind roll diameter [Øfinal] to 166 mm.

Diameter of uncoated drum [Ø] = 138 mm
Unwind roll diameter [Øfinal] = 166 mm
Correction factor [f] = ?

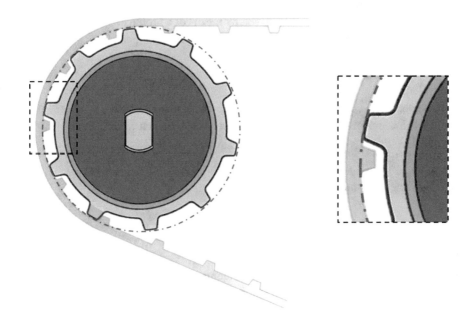

Fig. 5.16 Unwind roll diameter for positively driven thermoplastic belts

Formula:

$$f = \frac{\varnothing \text{final}}{\varnothing}$$

$$f = \frac{166\,\text{mm}}{138\,\text{mm}}$$

$$f = 1.2$$

In order to obtain the desired belt speed at the unwind roll diameter, one must select a drum motor in the catalogue, which is slower than the desired belt speed by the correction factor [f].

Example:
Desired belt speed [v] = 0.5 m/s
Correction factor [f] = 1.2
Drum motor catalogue speed vk = ?

Formula:

$$v_k = \frac{v}{f}$$

$$v_k = \frac{0.5\,\text{m/s}}{1.2}$$

$$v_k = 0.42\,\text{m/s}$$

In addition to the speed, the belt tension must also be calculated back to the drum motor catalogue value based on the unwind roll diameter.

Example:

Correction factor [f] = 1.2

Calculated belt pull [F] = 1611.12 N

Drum motor force – catalogue value [Fk] = ?

Formula:

$$F_k = F * f$$
$$F_k = 1611.12\,N * 1.2$$
$$F_k = 1933.34\,N$$

Appendix

Checklist for conveyor belt drive dimensioning

Question:	Source of information:
Is it a wet, damp or dry environment?	Conveyor manufacturer or conveyor belt operator
What is the minimum and maximum ambient temperature?	Conveyor manufacturer or conveyor belt operator
What is the altitude of the application above sea level?	Conveyor manufacturer or conveyor belt operator
What is the gradient of the conveyor belt? – Horizontal conveyor belt: slope = 0° – Straight ascending/descending conveyor: gradient [°] = ? – L-frame conveyor: gradient [°] and TL [mm] = ? – Gooseneck conveyor: gradient [°] and BL [mm] = ? – Z-frame conveyor: gradient [°], TL [mm] and BL [mm] = ?	Conveyor manufacturer or conveyor belt operator
How wide is the conveyor belt (BW) [m]?	Conveyor manufacturer or conveyor belt operator
How long is the conveyor from axis to axis (L) [m]?	Conveyor manufacturer or conveyor belt operator
What is the mass to be conveyed (m) [kg]?	Conveyor manufacturer or conveyor belt operator
What is the belt speed (v) [m/s]?	Conveyor manufacturer or conveyor belt operator
Are there any places where additional friction can occur (Fadd) [N]? For example: – Pivot points for L-frame, gooseneck or Z-frame conveyors *(approx. value: 50–100 N)* – Cleaner/scraper *(approx. value: 75–100 N)* – Brushes *(approx. value: 50 N)* – Accumulation conveying	Conveyor manufacturer or conveyor belt operator

© Springer-Verlag GmbH Germany, part of Springer Nature 2020
S. Hamacher, *The Drum Motor*,
https://doi.org/10.1007/978-3-662-59298-4

Question:	Source of information:
Which type of belt should be used? – Friction driven belt – Positively driven modular belt – Positively driven thermoplastic belt	Conveyor manufacturer or conveyor belt operator
Important belt data: – Belt specific mass (mb) [kg/m^2] – Belt thickness (Bt) [mm] – K1% value (K1%) [N/mm] *(for friction driven belts)*	Belt data sheet
What is the belt elongation (ε) [%]? (for friction driven belts)	Measurement on the belt surface, conveyor manufacturer or conveyor belt operator
What is the conveyor belt unwind roll diameter? – Lagging thickness (R) [mm] *(for friction driven belts)* – Pitch circle diameter (PCD) [mm] *(for modular belts)* – Profile outer diameter (OD) [mm] *(for thermoplastic belts)*	Calculation, conveyor manufacturer or conveyor belt operator
Friction factor between carrying surface and conveyor belt?	Table 5.1, conveyor manufacturer or conveyor belt operator

Formulary, calculations and rules

Symbol	Unit	Description:
α	°	gradient angle straight conveyor belt
α2	°	gradient angle for L-frame, gooseneck and Z-frame conveyors
Be	mm	distance between markings, tensioned conveyor belt
Be$_0$	mm	distance between markings, unstretched conveyor belt
BL	mm	bottom horizontal conveyor length, L and Z-frame conveyors
Bt	mm	conveyor belt thickness
BW	m	conveyor belt width (in meters [m])
CL	mm	conveyor length with gradient
ε	%	belt elongation
F	N	belt pull force at the unwind roll diameter
f	–	correction factor for drum motor catalogue values
f$_{alt}$	–	reserve factor higher than 1000 m above sea level
F$_{alt}$	N	belt pull for applications higher than 1000 m above sea level
F$_k$	N	belt pull converted to catalogue value
F$_{res}$	N	belt pull at unwind roll diameter plus reserve
g	N/kg	gravitational acceleration
h	mm	travel height
K1%	N/mm	force per millimetre of belt width at 1% belt elongation

L	m	axis-to-axis length of the conveyor
m	kg	mass to be conveyed
mb	kg/m^2	conveyor belt specific mass per m^2
mbu	kg	conveyor belt mass in upper strand
OD	mm	outer diameter, profile for thermoplastic belts
Ø	mm	outer diameter of uncoated drum motor
Øfinal	mm	final conveyor belt unwind roll diameter
P_{alt}	W	mechanical motor power higher than 1000 m above sea level
PCD	mm	pitch circle diameter of a modular belt
P_{mech}	W	mechanical motor power
R	mm	coating thickness for friction-driven conveyor belts
TE	N	belt tension force
TL	mm	top horizontal length for gooseneck and Z-frame conveyors
μ	–	friction factor between conveyor belt and carrying surface
v	m/s	desired belt speed
v_k	m/s	belt speed converted to catalogue value

Formulary and calculation method:

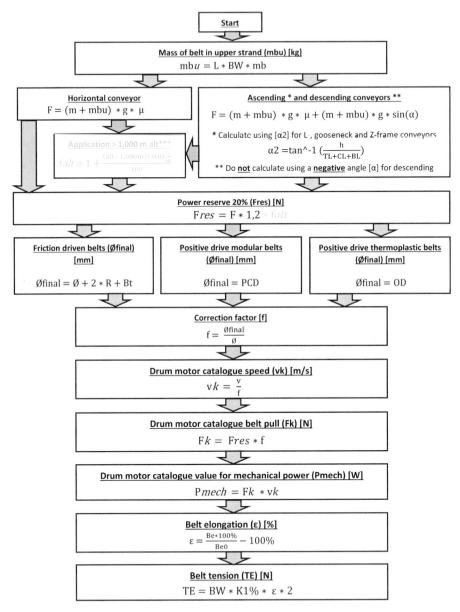

Start

Mass of belt in upper strand (mbu) [kg]
$$mbu = L * BW * mb$$

Horizontal conveyor
$$F = (m + mbu) * g * \mu$$

Ascending * and descending conveyors *
$$F = (m + mbu) * g * \mu + (m + mbu) * g * \sin(\alpha)$$
* Calculate using [α2] for L, gooseneck and Z-frame conveyors
$$\alpha2 = \tan^{-1}\left(\frac{h}{TL+CL+BL}\right)$$
** Do **not** calculate using a **negative** angle [α] for descending

Application > 1,000 m alt***
$$falt = 1 + \frac{(alt - 1.000m) * 0.01\frac{\%}{m}}{100}$$

Power reserve 20% (Fres) [N]
$$Fres = F * 1,2 * falt$$

Friction driven belts (Øfinal) [mm]
$$\text{Øfinal} = \emptyset + 2 * R + Bt$$

Positive drive modular belts (Øfinal) [mm]
$$\text{Øfinal} = PCD$$

Positive drive thermoplastic belts (Øfinal) [mm]
$$\text{Øfinal} = OD$$

Correction factor [f]
$$f = \frac{\text{Øfinal}}{\emptyset}$$

Drum motor catalogue speed (vk) [m/s]
$$vk = \frac{v}{f}$$

Drum motor catalogue belt pull (Fk) [N]
$$Fk = Fres * f$$

Drum motor catalogue value for mechanical power (Pmech) [W]
$$Pmech = Fk * vk$$

Belt elongation (ε) [%]
$$\varepsilon = \frac{Be * 100\%}{Be0} - 100\%$$

Belt tension (TE) [N]
$$TE = BW * K1\% * \varepsilon * 2$$

*** only for applications over 1,000 m

Evaluation:

The belt speed, converted to catalogue value (v_k) [m/s], must be as close as possible to one of the speeds specified in the catalogue.

The belt pull, converted to catalogue value (F_k) [N], must be less than or equal to the belt pull value specified in the drum motor catalogue.

The belt tension (TE) [N] must be less than or equal to the value specified in the drum motor catalogue for the maximum allowable belt tension.

The mechanical motor power (P_{mech}) [W] allows for better orientation in the catalogue, as drum motors are usually sorted by power. It is important that the values of (F_k) and (v_k) match the application.

Where possible, the **asynchronous drum motor winding** should be selected in the following order with respect to the number of poles:

1. **4**-pole
2. **2**-pole
3. **6**-pole
4. **8**-pole
5. **12**-pole

A suitable VFD must be selected for **synchronous drum motors**. The VFD must be able to control permanent magnet synchronous motors sensorlessly.

The output power of the VFD must match the motor power.

Execution and material selection:

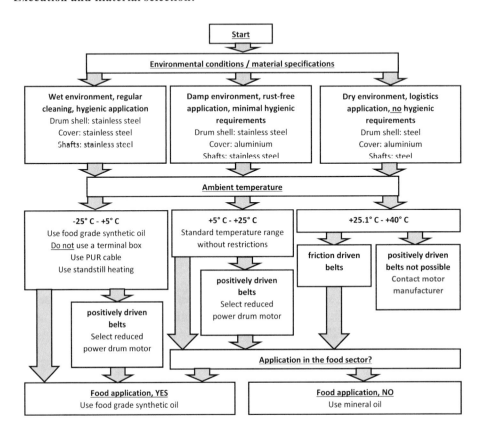